Engineering Aviation Security Environments—Reduction of False Alarms in Computed Tomography-Based Screening of Checked Baggage

I0879199

Committee on Engineering Aviation Security Environments—
False Positives from Explosive Detection Systems

National Materials and Manufacturing Board

Division on Engineering and Physical Sciences

NATIONAL RESEARCH COUNCIL
OF THE NATIONAL ACADEMIES

THE NATIONAL ACADEMIES PRESS
Washington, D.C.
www.nap.edu

THE NATIONAL ACADEMIES PRESS 500 FIFTH STREET, NW WASHINGTON, DC 20001

NOTICE: The project that is the subject of this report was approved by the Governing Board of the National Research Council, whose members are drawn from the councils of the National Academy of Sciences, the National Academy of Engineering, and the Institute of Medicine. The members of the committee responsible for the report were chosen for their special competences and with regard for appropriate balance.

This study was supported by the Department of Homeland Security Cooperative Agreement 04-G-045 with the National Academy of Sciences, Award 0840104 from the National Science Foundation, and Award DE-FG02-08ER46534 from the Department of Energy. Any opinions, findings, conclusions, or recommendations expressed in this publication are those of the authors and do not necessarily reflect the views of the organizations or agencies that provided support for the project.

International Standard Book Number- 13: 978-0-309-21479-7
International Standard Book Number- 10: 0-309-21479-3

This report is available in limited quantities from

National Materials and Manufacturing Board
500 Fifth Street, NW
Washington, DC 20001
nmmb@nas.edu
http://www.national-academies.org/nmmb/

Additional copies of this report are available from the National Academies Press, 500 Fifth Street, NW, Keck 360, Washington, DC 20001; (800) 624-6242 or (202) 334-3313; http://www.nap.edu.

Printed in the United States of America

THE NATIONAL ACADEMIES

Advisers to the Nation on Science, Engineering, and Medicine

The **National Academy of Sciences** is a private, nonprofit, self-perpetuating society of distinguished scholars engaged in scientific and engineering research, dedicated to the furtherance of science and technology and to their use for the general welfare. Upon the authority of the charter granted to it by the Congress in 1863, the Academy has a mandate that requires it to advise the federal government on scientific and technical matters. Dr. Ralph J. Cicerone is president of the National Academy of Sciences.

The **National Academy of Engineering** was established in 1964, under the charter of the National Academy of Sciences, as a parallel organization of outstanding engineers. It is autonomous in its administration and in the selection of its members, sharing with the National Academy of Sciences the responsibility for advising the federal government. The National Academy of Engineering also sponsors engineering programs aimed at meeting national needs, encourages education and research, and recognizes the superior achievements of engineers. Dr. C.D. (Dan) Mote, Jr., is president of the National Academy of Engineering.

The **Institute of Medicine** was established in 1970 by the National Academy of Sciences to secure the services of eminent members of appropriate professions in the examination of policy matters pertaining to the health of the public. The Institute acts under the responsibility given to the National Academy of Sciences by its congressional charter to be an adviser to the federal government and, upon its own initiative, to identify issues of medical care, research, and education. Dr. Harvey V. Fineberg is president of the Institute of Medicine.

The **National Research Council** was organized by the National Academy of Sciences in 1916 to associate the broad community of science and technology with the Academy's purposes of furthering knowledge and advising the federal government. Functioning in accordance with general policies determined by the Academy, the Council has become the principal operating agency of both the National Academy of Sciences and the National Academy of Engineering in providing services to the government, the public, and the scientific and engineering communities. The Council is administered jointly by both Academies and the Institute of Medicine. Dr. Ralph J. Cicerone and Dr. C.D. (Dan) Mote, Jr., are chair and vice chair, respectively, of the National Research Council.

www.national-academies.org

COMMITTEE ON ENGINEERING AVIATION SECURITY ENVIRONMENTS—FALSE POSITIVES FROM EXPLOSIVE DETECTION SYSTEMS

SANDRA HYLAND, BAE Systems, *Chair*
CHERYL BITNER, Pioneer UAV, Inc.
R. GRAHAM COOKS, Purdue University
CARL R. CRAWFORD, Csuptwo, LLC
B. JOHN GARRICK, Executive Consultant, Laguna Beach, Calif.
CONSTANTINE GATSONIS, Brown University
GARY H. GLOVER, Stanford University
SUBHASH R. LELE, University of Alberta
HARRY E. MARTZ, JR., Lawrence Livermore National Laboratory
WILLIAM Q. MEEKER, Iowa State University

Staff

EMILY ANN MEYER, Study Director
TERI THOROWGOOD, Administrative Coordinator (until December 2009)
LAURA TOTH, Program Assistant
RICKY D. WASHINGTON, Executive Assistant

Preface

The face of aviation security changed drastically in the wake of the terrorist attacks on the United States on September 11, 2001. Among the changes was the requirement, mandated by the Aviation and Transportation Security Act of 2001,[1] that as of December 31, 2003, all checked baggage on U.S. flights be scanned by explosive detection systems (EDSs) for the presence of any potential explosives threat.

In most airports, this scanning is performed by a computed tomography (CT)-based device. Such devices are based on the same technology as that used for medical CT scanners, with minor modifications so that the scanners can perform in the significantly larger scale of operation required in airports. Medical scanners perform well in a clinical setting; however, modifying them to scan for explosives in an airport setting can result in shortcomings—including those related to reliability and the false alarm rate—owing to the very different scale of operation and the resulting greater demands on the equipment.

The Committee on Engineering Aviation Security Environments—False Positives from Explosive Detection Systems addresses some of these issues related to reliability and makes recommendations for research and administrative directions that may allow for a reduction in the false alarm rate. Throughout the study process, the committee balanced considerations related to a reduction in false alarms with concerns about increased risk of missed detection.

The committee acknowledges with thanks those who spoke at meetings. The committee is also grateful for the support of National Research Council staff throughout this project.

Sandra Hyland, *Chair*
Committee on Engineering Aviation
Security Environments—False Positives from
Explosive Detection Systems

[1] Public Law 107-71, signed into law November 19, 2001.

Acknowledgment of Reviewers

This report has been reviewed in draft form by individuals chosen for their diverse perspectives and technical expertise, in accordance with procedures approved by the National Research Council's (NRC's) Report Review Committee. The purpose of this independent review is to provide candid and critical comments that will assist the institution in making its published report as sound as possible and to ensure that the report meets institutional standards for objectivity, evidence, and responsiveness to the study charge. The review comments and draft manuscript remain confidential to protect the integrity of the deliberative process. We wish to thank the following individuals for their review of this report:

Peter Edic, GE Global Research,
Maryellen L. Giger, University of Chicago,
Peter Hovey, University of Dayton,
Najmedin Meshkati, University of Southern California,
Stephen Pollock, University of Michigan,
Elan Scheinman, Reveal Imaging Technologies, and
Ron Willey, Northeastern University.

Although the reviewers listed above have provided many constructive comments and suggestions, they were not asked to endorse the conclusions or recommendations, nor did they see the final draft of the report before its release. The review of this report was overseen by Hyla S. Napadensky, Napadensky Energetics, Inc. Appointed by the NRC, she was responsible for making certain that an independent examination of this report was carried out in accordance with institutional procedures and that all review comments were carefully considered. Responsibility for the final content of this report rests entirely with the authoring committee and the institution.

Contents

APPENDIXES

Summary and Recommendations

The Aviation and Transportation Security Act (Public Law 170-71) of November 19, 2001, mandated that as of December 31, 2002, all checked baggage on U.S. flights be scanned by explosive detection systems (EDSs) for the presence of potential explosives threats. In response, the Department of Homeland Security's (DHS's) Transportation Security Administration (TSA) embarked on a program to quickly procure and deploy certified EDS equipment at all U.S. airports. Although any TSA-certified method of detecting explosives will meet the requirements of the Aviation and Transportation Security Act, the requirement is met in most airports through x-ray computed tomography (CT)-based systems. Now that CT-based detection systems have been in use for more than 10 years, TSA seeks to improve the performance of its baggage screening systems through such measures as better detection algorithms and more effective EDS equipment—especially to reduce the number of false alarms and thereby reduce the costs of screening checked baggage.

This report, from the National Research Council's Committee on Engineering Aviation Security Environments—False Positives from Explosive Detection Systems, examines potential technical enhancements, opportunities to foster innovation, and data requirements for reducing the false alarm rate (the committee's full statement of task appears in Appendix E). This summary provides a brief overview of the report along with the committee's key conclusions, findings, and recommendations. The supporting discussion appears in the chapters that follow, along with additional conclusions, findings, and recommendations as appropriate to the topic of discussion.

EXPLOSIVES DETECTION USING COMPUTED TOMOGRAPHY

CT does not directly detect explosives. Rather, CT is used in combination with automated threat recognition (ATR) algorithms to identify objects whose properties fall within specified ranges of the properties of explosives. The process starts with the CT scanner emitting x-rays that pass through a bag. X-ray detectors convert the received x-ray flux to electrical signals, which are then processed to reconstruct a series of cross-sectional images of the bag, as well as estimates of atomic density (atoms per unit volume), mass, and size and shape of items within the bag.[1] These cross-sectional images are then analyzed by an ATR algorithm that uses information in the images to determine whether the images satisfy a set of criteria consistent with the bag containing an object that is an explosives threat. The ATR algorithms have been developed and refined over many years to alert on threat amounts of materials that fall within a specified density and mass range ("detection window"). If these criteria are met, then the object is declared a "potential threat" and an alarm is raised.

There are four possible results of such a screening, as shown in Figure S-1:

- A threat is present and the system alarms, resulting in a true detection.
- A threat is present and no alarm is raised, resulting in a missed detection.
- No threat is present and the system alarms, resulting in a false positive.
- No threat is present and no alarm is raised, resulting in a true negative.

[1] Some vendors may also use the filtered back-projection data and the images from the digital radiography line scanner in their ATR algorithm.

	Alarm	No Alarm
Threat Present	True Detection	Missed Detection
No Threat Present	False Positive	True Negative

FIGURE S-1 A contingency table of the potential results of an interrogation of checked baggage by a computed tomography-based explosive detection system.

Because other, non-threat items can also fall within these specified ranges, there will always be a greater-than-zero chance of false alarm as long as there is also a greater-than-zero probability of detection. The difficulty in isolating threat materials from non-threat materials can be seen in Figure S-2, which shows typical density ranges for some threat and non-threat materials and the overlapping densities of some materials. It is inevitable then that relying solely on density and mass to identify threat materials will lead to some misidentification and false alarms.

Human screeners must resolve alarms raised by the EDSs. First, following a TSA-established "on-screen alarm-resolution protocol," a screener at the baggage viewing station is presented with information from the scan (such as cross-sectional images of the bag and specifics of any suspicious objects generated by the ATR algorithm), which is used to either clear the alarm or to send the bag for further inspection. Bags that cannot be cleared are handled according to local regulations for potential explosive threats.

IMPLICATIONS OF A FALSE POSITIVE RATE

The TSA estimates that each percentage point of the current false alarm rate costs the government tens of millions of dollars per year. The main element of these costs for resolving false alarms is the total number of personnel required to screen baggage in U.S. airports, because every bag that causes an alarm must be sent for further inspection. However, there are other elements that contribute to the cost of resolving these alarms—including those associated with the infrastructure for segregating bags for manual inspection, with maintaining controlled areas for opening bags, and with tracking bags. There is also a cost in time and convenience to travelers who must arrive earlier at airports to ensure baggage can be screened in time for their flight's departure.

In addition to the added expense that it generates, the process of resolving false alarms may increase the net risk to air transport because the time and personnel allocated to resolving false alarms may take away from other security efforts. Moreover, some studies suggest that the current high false alarm rate may in fact reduce the likelihood of identifying an actual threat, as screeners have come to expect that the cause of an alarm is a non-threat.[2]

[2] See, for example, Mathias S. Fleck and Stephen R. Mitroff, Rare targets are rarely missed in correctable search, Psychological Science 18:943-947, 2007; and Anina N. Rich, Melina A. Kunar, Michael J. Van Wert, Barbara Hidalgo-Sotelo, Todd S. Horowiths, and Jeremy M. Wolfe, Why do we miss rare targets? Exploring the boundaries of the low prevalence effect, Journal of Vision 8:1-17, 2008.

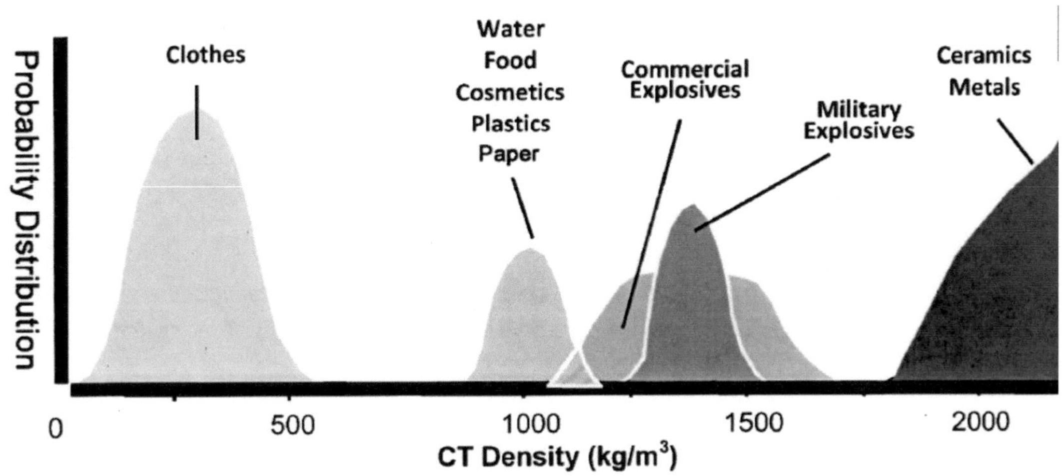

FIGURE S-2 Notional distribution of threats and non-threats in computed tomography (CT) density space. Clothes are clearly distinguishable as a single-valued function of density, but other non-threat materials commonly found in passenger bags show some overlap in material density with some threat materials.

Adding to the risks described above associated with personnel being diverted from detecting other threats and screeners expecting non-threats is the fact that as currently deployed, CT-based EDSs are tuned only to detect a certain set of explosives. Reducing the false alarm rate without increasing the rate of false negatives would free up resources to develop and deploy capabilities for identifying explosive materials.[3]

TECHNICAL APPROACHES TO REDUCING THE FALSE POSITIVE RATE

Although there is no way to completely eliminate all false alarms, there are a number of potential technical approaches that would improve the false positive rate. Each offers some potential improvements in system performance while imposing additional development or operational costs—and some also carry the risk of decreasing the detection rate.

- *Adjust the operating point on the receiver operating characteristic curve.* One way to characterize the performance of a detector system is a plot of threat identification (probability of detection, or PD) versus misidentification of an innocuous item as a threat (probability of false alarm, or PFA), known as the receiver operating characteristic (ROC) curve, as shown in Figure S-3. By moving along the curve and narrowing the detection windows in order to eliminate the misidentification of non-threat materials there will be a corresponding risk of decreasing the detection rate and missing a true threat. Moving along the curve in the other direction and expanding the detection window to ensure the capture of all threat materials will result in capturing non-threat materials and increasing the false alarm rate.
- *Improve image processing.* The correction step in CT image reconstruction uses software to compensate for imperfections in the projection data acquired during scanning, but these corrections are not perfect, and image artifacts will always remain. These artifacts increase uncertainty in the measurement and evaluation of the objects that are being scanned, and can result in such things as mis-estimation of object mass, a widening of density windows, and inaccurate region building (which leads to

[3] Studies addressing the detection of novel and liquid explosives are discussed in the subsection entitled "Dual-Energy Scanning" in Chapter 2 of this report.

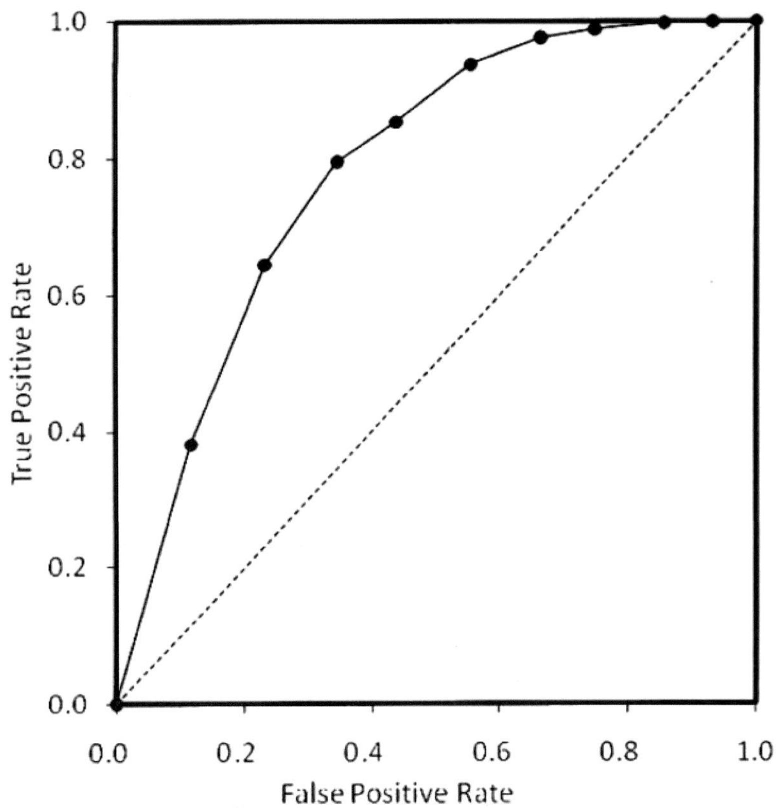

FIGURE S-3 An example of a receiver operating characteristic curve.

over-aggregation of different objects). The net effect is to lower confidence in the estimated characteristics of an object within a bag, forcing the threat-defining windows to be widened, which results in a concurrent increase in false alarms. Thus, improvements in the image reconstruction and correction process could lead to a lower false alarm rate.

- *Slow the bag-processing speed.* More scanning time allows for more detailed scanning. Indeed, with unlimited time to screen a bag, the estimations of mass and density would be substantially improved. The tradeoff that must be considered is whether increases in automated bag-screening time can be justified by the resulting decrease in the number of false alarms.
- *Perform additional scans.* One way to reduce the probability of false alarms and to improve the probability of detection of CT-based EDSs is to increase the number of cross-sections that the machine takes of an object. More cross-sections usually lead to a better probability of correct discrimination in recognizing whether an object is a threat or a non-threat. The number of cross-sections that a machine takes can be increased either by changing the current hardware or by passing a bag through existing CT scanners multiple times in such a manner that the bag is positioned somewhat differently for each scan. Of course, the costs associated with implementing rescanning, such as increased screening time, additional routing hardware, and modifications to scanners, must be justified by a sufficient decrease in false alarms.
- *Investigate ways to better distinguish between materials' similar density.* Adding atomic number to the screening criteria via dual-energy CT technology has the potential to improve an ATR algorithm's ability to distinguish between threat and non-threat materials of the same density, and thus lower the probability that the EDS would give a false alarm. Some researchers have found that extensive

calculations would be required to clear a whole bag;[4] more than a minute may be needed even if a graphical processing unit is used to accelerate the computation.[5] Further exploration is needed to obtain a full understanding of its advantages and limitations.

- *Supplement computed tomography screening with additional technologies.* Supplementing the decision from the ATD with information from another technology is another way to reduce the need for manual inspection of baggage. Such information might come from a different form of imaging technology, such as x-ray diffraction; a chemical analysis, such as mass spectrometry; or data from other sources, such as carry-on-baggage and passenger-screening checkpoints, perimeter-surveillance data, or even information about passengers' behavior or travel habits. Coupled with CT, this information might be used to reduce the overall false alarm rate without increasing the risk for false negatives.

Finding: Based on the information available at this time about the performance characteristics of these approaches and available data on the actual sources of false alarms raised by today's explosives detection systems, it is not possible to establish which are most promising or merit significant investment.

For one of these approaches, adjusting the operating point on the receiver operating characteristic curve, the committee has a specific recommendation concerning avenues for further investigation.

Recommendation: The Transportation Security Administration, through the Transportation Security Laboratory, should support human-factor studies to assess the impact on overall system performance, that is, the EDS plus the screener resolution, when the operating point on the explosive detection system's receiver operating characteristic curve is adjusted so that both the probability of detection and probability of false alarm are lowered. If the results of such studies determine that screener attention is degraded by the expectation that every alarm is a false alarm, the TSA should consider implementing adjustments to the operating point on the receiver operating characteristic curve and allowing vendors to reduce probability of detection in an airport setting to the minimum rate required for certification.

INCENTIVIZING AND ENABLING INNOVATION AND IMPROVEMENT

EDS vendors told the committee that the TSA provides them with few incentives to improve the performance of their equipment. Additionally, although TSA aims to improve the false alarm performance of EDSs for baggage screening, the committee was made aware of no clear plan from the TSA to implement improvements in the performance of fielded systems. Without changes to current TSA policy, there will be insufficient incentives for vendors to spend money to develop improvements beyond the necessary fixes for known problems.

Creating incentives for vendors and the technical community to develop improvements will require an organizational framework that includes a known path for the deployment of candidate improved technologies, a realistic strategy for fielding proven improvements, and specific incentives for vendors to provide equipment that performs better than would be necessary to meet baseline requirements. The committee believes that the DHS and the TSA, in cooperation with the equipment vendors, should develop a realistic, long-term strategy for the performance improvement of EDS equipment in an airport setting.

---

[4] Wenyuan Bi, Zhiquiang Chen, Li Zhang, and Yuxiang Xing, A volumetric object detection framework with dual-energy CT, pp. 1289-1291 in IEEE Nuclear Science Symposium Conference Record, IEEE Piscataway, N.J., 2008.

[5] Guori Yan, Jie Tian, Shouping Zhu, et al., Fast cone-beam CT image reconstruction using GPU hardware, Journal of X-Ray Science and Technology 16(4):225-234, 2008.

Although priorities in a long-term plan involving EDS equipment would necessarily change on the basis of changing threat environments and other outside influences, a long-term plan developed cooperatively would allow companies to evaluate their risk-and-reward strategy in a more stable investment environment.

One possibility is adopting a different contracting mechanism. Performance-based logistics (PBL) has been successfully used by the Department of Defense in somewhat analogous circumstances. Under a PBL-based contract, the government and the equipment vendor work together to determine key performance indicators for the equipment, and the government provides incentives for the vendors to invest in improvements with a reasonable expectation that these improvements will be evaluated and implemented if successful.

Recommendation: In order to better capitalize on improvements and provide vendors with the necessary incentives to invest in research that will lead to better performance metrics, the TSA should consider adoption of a different contract structure for the procurement and maintenance of the computed tomography-based explosive detection systems used for checked baggage, as well as for other screening technologies. One approach worth considering is performance-based logistics contracting, which is currently used by the Department of Defense.

Evaluating Proposed Vendor Enhancements

The committee also heard frustration from vendors regarding the prospects that improvements they develop will be purchased and fielded by the TSA. Each vendor that the committee heard from[6] described improvements that could be fielded now but that were being hindered by TSA testing requirements (such as those not permitting candidate improved algorithms to be tested without putting the entire system through the certification process) or by a lack of guidance on how changes were to be evaluated or implemented. Companies will invest in technology improvements that can reasonably be expected to generate a return on the investment. Procurement cycles need to be structured such that vendors will be willing to make appropriate investments in better performance.

Of course, not every suggested change will merit fielding. Rather, TSA will need to create a framework to evaluate suggested enhancements. A first step in that process could be the establishment of a "technology board" composed of individuals knowledgeable about the technology and with broad experience in the technology, testing, and field requirements. The group's charter would be to evaluate proposed technology improvements, to identify evaluation methods, to assess the outcomes of tests, and to identify necessary process changes.

Conclusion: The TSA lacks a structured plan for implementing improved EDSs that would give vendors an opportunity to plan research funding and priorities in accordance with the TSA plan.

Cooperation with University Researchers and Other Outside Industries

Researchers at universities, government laboratories, and other industrial companies, funded from both private and government sources, have long been working in image reconstruction and processing and ATR algorithms. However, such research is not currently being conducted in coordination with TSA or any of the EDS vendors, and the committee saw no structure in place for such researchers to partner with either the government or EDS vendors for the development, evaluation, or fielding of improvements.

---

[6] Representative of General Electric (GE) Security and of L-3 Communications, presentations to the committee, February 12, 2009, San Francisco, California.

Conclusion: The engagement of more members of the academic and industrial communities, as well as of those in the medical diagnostics and military communities having theoretical and applied expertise in image reconstruction and target recognition, could lead to increases in the effectiveness (and, in particular, decreases in false alarms) of CT-based explosives detection.

Recommendation: The TSA should develop a plan to provide appropriate incentives not only for EDS vendors but also for third parties and researchers in academia in order to improve the overall performance of computed tomography-based EDSs, including their rates of false alarms. Incentives should be provided for both short- and longer-term improvements.

Decoupling Image Acquisition from Post-Processing to Foster Innovation

Many comparisons can be drawn between CT-based EDS and medical CT. One of the biggest differences is that whereas CT-based EDSs had to be deployed almost universally in a very short timeframe, the development of medical CT scanners has occurred over a period of many years, driven by a broad range of requirements. Further, unlike CT-based EDS where there are only a few vendors and a single customer, the vendors of medical CT scanners compete in a broad market on image quality, device flexibility, and cost, and maintain extensive research and development efforts to remain competitive.

Medical CT also enjoys an open environment and standardized image format (DICOM), which has opened the door to academic participation in post-processing innovations in three- and four-dimensional visualization and computer-aided diagnosis programs[7] because details of the scanner process were separated from those details related to the processing of the images for specific applications. This separation allowed the scanner vendors to retain control over propriety details of the acquisition of images, but it provided easy access to academic and industrial researchers. Based on the positive experience with DICOM, there is now a move toward standardizing the image format of CT used for EDSs.

Finding: The introduction of an industry-standard medical image format (DICOM) in 1993 fostered the development of a diverse and innovative array of diagnostic and therapeutic image visualization, processing, and automated detection/diagnostic products, fueled by the panoply of academic and private-sector research laboratories with extensive experience in the field.

Recommendation: The Department of Homeland Security should promote the rapid acceptance of a standardized format for EDS images for all TSA-certified machines.

Separating the acquisition of CT images from the post-processing programs will help enable greater competition for the development of the post-processing programs. Broader participation by these highly experienced groups with diverse backgrounds in image processing would make it likely, the committee believes, that new methods would be developed that may improve the detection and classification efficiency of baggage scanners. It should be noted, however, that although opening up post-processing to a wider community may lead to useful advances, it does not address the quality and completeness of the images provided by the image acquisition and reconstruction stages.

---

[7] See, for example, "SecurView Diagnostic Workstations," available at http://www.hologic.com/en/breast-imaging/diagnostic-workstations/, accessed September 12, 2010; and Fang-Fang Yin, Maryellen L. Giger, Kunio Doi, Charles E. Metz, Carl J. Vyborny, and Robert A. Schmidt, Computerized detection of masses in digital mammograms: Analysis of bilateral subtraction images, Medical Physics 18(5, September):955-963, 1991.

Overarching Advice

The committee's overarching recommendation concerning innovation and improvement is as follows.

Recommendation: The TSA should develop a long-term strategy for the continuous improvement of performance. Involving all interested parties including EDS vendors and users would increase the probability that all stakeholders work toward the same goals.

In addition to specific technology improvements, there are a number of areas where TSA might better incentivize or foster innovation—providing incentives for contractors, better defining mechanisms for implementing contractors' improvements, fostering cooperation with universities and outside researchers, and promoting the decoupling of the image acquisition process from the post-processing algorithm.

CERTIFICATION TESTING AT THE TRANSPORTATION SECURITY LABORATORY IS USEFUL BUT DOES NOT REFLECT REAL-WORLD CONDITIONS

Certification testing of EDSs and subsequent performance testing in an airport setting is one source of information on EDS performance and causes of false alarms. To be certified, a machine must demonstrate the ability to detect a number of categories of explosives, with each category having a specific detection threshold (i.e., level of detection that must be met). The machine must also meet an average detection threshold across all categories of explosives and not exceed a maximum false alarm rate (which is tested separately from the detection).

TSA's Transportation Security Laboratory uses two sets of bags for certification tests. One set contains one threat per bag and the other has no threats. The "threat" test set is, of course, not fully representative of the bags likely to be found in an airport setting—which may contain multiple potential threats. As a result, the probability of detection established for an EDS at the TSL may not be maintained in an airport setting. However, the use of bags more representative of those seen in an airport setting would be quite complicated, given the potential variations in bags and contents. Additionally, the use of a simple, well-defined set of test bags has the advantage of allowing all manufacturers to be tested against a common standard.

The certification testing also does not account for the humans in the screening process loop. In an airport setting, there can, for example, be pressure to clear bags in order to make delivery deadlines and variability in the way in which the on-screen alarm-resolution protocol is carried out.[8] Beyond these limitations, current testing does not address the quality of the CT scanner image data output and analysis output once scanners have been deployed. Nor is information about false alarms analyzed and fed back into future iterations of the ATR software.

Conclusion: Certification testing at the Transportation Security Laboratory fills a specific and useful role. However, it should not be used as the sole basis for predictions of performance in an airport setting.

Recommendation: The TSA should develop procedures for periodic verification to ensure that fielded EDSs meet detection-performance-level standards that correspond to the requirements for EDS certification. In addition to monitoring detection capability directly (e.g., using standard bag sets and red-

---

[8] See, for example, Sara Kraemer, Pascale Carayon, and Thomas F. Sanquist, Human and organizational factors in security screening and inspection systems: Conceptual framework and key research needs, Cognition, Technology, and Work 11:29-41, 2009.

team testing), these procedures should include the frequent monitoring of critical system parameters (e.g., voltages and currents) and imaging parameters (e.g., image resolution and image noise) to detect system problems as soon as they arise. For purposes of monitoring EDS performance, the TSA and EDS vendors should develop specification limits for all critical system parameters (and their tolerances) that could be monitored frequently and recorded to track changes in performance during normal operations or to verify performance after maintenance or upgrading.

USE A DATA-DRIVEN APPROACH TO REDUCE FALSE POSITIVES

A detailed, quantitative understanding of the root causes of false positives is important if the TSA is to reduce the costs associated with these false positives without increasing other risks. For instance, the overall false alarm rate includes two distinct "populations" of bags, each of which would require a different approach to reducing false alarm rates:

- The first population includes bags for which the EDS cannot make a decision—so-called "exceptions," such as bags containing solid objects that cannot be penetrated by the EDS x-rays, mis-tracked bags, and bags that are poorly positioned in the EDS in such a way that the EDS cannot interrogate the entire bag ("cut bags"). These exceptions are sent directly to the baggage inspection room without the opportunity for a screener to evaluate the image and clear the bag.
- The second population includes bags whose contents include items that are misidentified by the EDS as potential threat items—for example, when the item's properties fall within the window defined for threat items, or multiple items are mistakenly aggregated into a single object that meets the criteria for a potential threat item.

Without systematic data that can be used to establish how much each population of bags contributes to the overall false alarm rate, or what the specific causes of false alarms are within each population, it is difficult to know what the right course of action is.

Recommendation: The Transportation Security Administration should track broad categories of bags with the goal of understanding how each category contributes to the cost of resolving false alarms.

Categories should include the following: the number of bags scanned, the number of bags declared exceptions, the number of bags declared potential threats by the EDS and cleared by the screener using the on-screen alarm-resolution protocol, and the number of bags declared potential threats by the EDS and sent by the screener to the baggage-inspection room for further inspection. Tracking these data over multiple airports and multiple seasons would give the TSA a better overall understanding of the cost drivers contributing to the false alarm rate.

Also, it is possible that the wide range of false-positive images that current screening practices detect could be usefully partitioned into a manageable number of classes of baggage items (e.g., cosmetics, food-stuffs, or metal) and non-bag related causes (e.g., algorithm issues, losing track of bags during the screening process, or hardware faults). Again, better data from the entire screening process is needed to assess the merit of this approach.

In short, better data would go a long way toward improving TSA's understanding of the causes of false alarms and allow for a more structured approach toward reducing them. Put another way, without a better understanding of how well (or poorly) the systems are working, it is difficult to make improvements.

The relevant data required to answer such questions includes information about false positives (including data captured by the EDS machines themselves, data about on-screen alarm resolution, and data about resolution through manual screening) and information about false negatives (including results of red-team testing and actual adverse events).

The EDS machines are already equipped to generate useful data for better understanding false positives. To be certified as an EDS, a machine is required to have the capability of recording data about the bags being scanned, including the potential threats identified by the ATR algorithm. However, there is no requirement for either the TSA or the EDS vendors to store or analyze these data. Although vendors have collected some data in the context of particular studies for TSA,[9] officials in the TSA with whom the committee spoke indicated that larger-scale data collection and analysis are not being done.

Also, the ATR and On Screen Alarm Resolution Protocol (OSARP) data alone do not provide any information about alarms that must be resolved through manual inspection. However, in its tour of the baggage inspection room at San Francisco International Airport, the committee saw no mechanism for collecting data on the results of these inspections, nor any systematic framework for such information—such as a categorization of causes of alarms. Indeed, the committee believes that San Francisco International Airport is representative of airports throughout the United States in this respect, an observation confirmed by a TSA official who briefed the committee subsequent to its visit.

Finding: The low prevalence of the true positives in an airport setting may make it nearly impossible to measure probability of detection with humans-in-the-loop without forcing true positives via red-team testing.

Finding: Discussion with TSA officials, airport personnel, and vendors indicates some limited-scale data collection and laboratory studies that have enabled the sources of false alarms to be broadly identified. However, system-wide data collection and analysis of the sort necessary to seek out the root causes and guide sustained improvements are not being done.

Conclusion: Without more systematic data on the rates and specific causes of false alarms, the TSA cannot determine what changes are likely to result in reduced false alarm rates and, in fact, does not have the infrastructure in place to determine if an implemented change would result in improved performance.

Recommendation: The TSA should develop a system for sharing false-positive data with detection-equipment vendors, including ATR algorithm developers and, when reasonable, with baggage vendors. Vendors should have a clear picture of how well or poorly their own equipment and that of their competitors is operating in an airport setting.

Recommendation: The TSA should develop a categorization system to record particular causes of false alarms for baggage sent to the baggage-inspection room. TSA should develop a database to store this information and use it to monitor performance variation and trends over time.

A system for collecting, managing, and providing access to this data should be put in place, along with capabilities for viewing, reporting, and analysis—as well as export for special studies such as quantitative risk assessments (QRAs), or anomaly detection (i.e., when a sudden change occurs in the "normal" behavior of an EDS in an airport setting). Commercial off-the-shelf software for building such systems is readily available and reasonably priced, although additional investment in hardware and training will still be necessary.

Recommendation: The TSA should employ risk assessment methods to obtain a better understanding of the causes of false positives at both the system and the component level.

---

[9] Representative from Reveal Imaging Technologies, Inc., presentation to the committee, April 29, 2009, Washington, D.C.

QRA could also be an effective approach to analyzing the probability of explosives in airline baggage and for assessing the effects that changes to the baggage-inspection system will have on both probability of false alarm and probability of detection.

Recommendation: The Transportation Security Administration should work with the Transportation Security Laboratory to collect and analyze field data in order to characterize the overall performance of the system by computing statistically valid estimates of probability of detection and probability of false alarm for today's CT-based EDSs. These analyses should also be used to better understand the sources of false positives by determining the dependence of these probabilities on material characteristics of potential explosives threats, the variability in the material characteristics, and the characteristics of non-threat materials typically contained in checked bags. These estimates should then be used as baselines for determining the ability of potential improvements to reduce false alarms.

Recommendation: In addition to collecting performance data on a routine basis, the TSA should, from time to time, conduct special studies and experiments for the purpose of obtaining additional information that would be useful for improving the baggage-inspection processes.

In light of the limited information available today, the committee recommends that the TSA limit its own spending on replacement equipment to allow for learning to inform future expenses.

Recommendation: The TSA should not fund an overall replacement of fielded explosive detection systems, because replacing all the units in service with currently available technology would not allow for learning in an airport setting to inform future performance improvements. Instead, the TSA should plan its capital spending for explosives detection improvements over a period of time sufficient to allow several generations of technology to be fielded on a limited basis, evaluated, and iteratively improved—thus leading to a gradual improvement in the overall field performance of CT-based explosives detection systems.

However, once the sources of false alarms are better understood, the investment made in the data collection and analysis has the potential to result in a high rate of return based on a targeted approach to false alarm reduction. Table S-1 outlines some of these options.

TABLE S-1 Potential Solutions to the False Positive Rate Based on Cause

Cause	Possible Solutions
Poor image quality (streaking, agglomeration, etc.)	Improvements to the post-processing algorithm Image standardization Slowing scan speed Additional scans
Algorithm sensitivity	Additional scans Adjusting operating point on the receiver operating characteristic curve
Overlap between threat and non-threat materials	Dual-energy scans Supplement with orthogonal technologies such as mass spectrometry or x-ray diffraction
Exceptions (cut bag, time-out)	Additional scans
Shield alarm	Supplement with orthogonal technologies such as mass spectrometry

1

Introduction

BACKGROUND AND REQUEST FOR STUDY

With the enactment of the Aviation and Transportation Security Act (Public Law 170-71) on November 19, 2001, the Transportation Security Administration (TSA) was created as a separate entity within the U.S. Department of Transportation.[1] The act also mandated that as of December 31, 2002, all checked baggage on U.S. flights be scanned by explosive detection systems (EDSs) for the presence of potential explosives threats. As a result of the need to deploy EDS equipment quickly and universally, the procurement and installation of certified systems were emphasized by the TSA, as was the development of alternate equipment and procedures to provide screening where certified equipment was not yet available. Although any TSA-certified method of detecting explosives will meet the requirements of the Aviation and Transportation Security Act, in most airports this is provided by computed tomography (CT)-based systems.

Although the system is called an explosive *detection* system, CT does not have the ability to detect explosives. Computed tomography was originally developed as a medical diagnostic tool that uses x-rays transmitted through the scanned area to provide a three-dimensional image of the human body. For aviation security applications, the transmitted x-rays are reconstructed into a three-dimensional image of the scanned bag, and software algorithms (see the subsection entitled "Automated Threat Recognition" in Chapter 2 of this report) are applied to that image to estimate the material properties of the items within the bag and to compare those estimated properties to a set of criteria defined by the TSA for explosive threat items. These criteria have been developed over many years to detect specified materials at levels that would pose a threat, but they are not perfect. No algorithm can always find a threat item and never misidentify an innocuous item as a threat. Baggage screening is always a trade-off between false alarms and missed detections (also called false negatives). The TSA has put a high priority on maintaining a high level of detection, and as a result it must deal with a concomitantly high level of false alarms.

To gain a better understanding of and to be better able to address the issue of false alarms in EDSs, the Department of Homeland Security Science and Technology Directorate (DHS S&T) through the TSA requested a study from the National Research Council (NRC) of the National Academies. As a result of this request, the NRC established the Committee on Engineering Aviation Security Environments—False Positives from Explosives Detection Systems.

The committee's full statement of task is as follows:

> An ad hoc committee will examine the technology of current aviation-security explosive-detection systems (EDSs) and the false positives produced by this equipment. In assessing methods to reduce the EDS false-alarm rate, the committee will:
>
> 1. Examine and evaluate the causes of false positives in aviation explosive-detection systems, including considering the role of equipment design standards that rely on the fusion of explosive density measurement, total mass, and shape effects.
> 2. Assess the impact false positive resolution has on personnel and resource allocation.

[1] The TSA was later moved to the Department of Homeland Security after its creation on November 25, 2002.

3. Make recommendations on mitigating false positives without increasing false negatives, considering both technology and personnel approaches and related short- and long-term research. The committee recommendations will also bear in mind any risk of increased missed detection.

Although the TSA recognized that false alarm rates for CT-based EDSs in an airport setting could be substantially higher than the false alarm rates measured in certification testing (see the section entitled "Testing at the Transportation Security Laboratory" in Chapter 2), the agency and policy makers believed that it was nevertheless important to field the equipment and provide screening of all checked baggage. Now that CT-Based EDSs have been employed for more than 10 years, the DHS S&T is focusing on better detection algorithms and more reliable equipment in order to reduce the number of false alarms and thereby reduce the costs of screening checked baggage.

The Transportation Security Administration estimates that each percentage point of the current false alarm rate costs the government millions of dollars per year. The main element of these costs for resolving false alarms is personnel time, because every bag that causes an alarm must be further inspected. However, there are other elements that contribute to the cost of resolving these alarms—including the infrastructure for segregating bags for manual inspection, controlled areas for opening bags, and tracking bags to the baggage-inspection room (BIR) and back to their designated flight.

In addition to the added expensed that it generates, the process of resolving false alarms may increase the risk to the air transport infrastructure because the time and personnel allocated to resolving false alarms may take away from security efforts that could detect or deter other threats. Moreover, some studies suggest that the current high false alarm rate may in fact reduce the likelihood of identifying an actual threat, as screeners have come to expect that the cause of an alarm is a non-threat.[2]

Adding to the risks described above associated with personnel being diverted from detecting other threats and screeners expecting non-threats is the fact that as currently deployed, CT-based EDSs are tuned only to detect certain explosives. Improving the algorithm and image-processing systems and reducing the false alarm rate to an acceptable level without increasing the rate of false negatives would clear the way for the work necessary to detect additional explosive materials.[3]

In discussions with the committee, the Transportation Security Administration and the sponsor at DHS S&T indicated that its primary interest was in the contributions to the false alarm rate made by the EDS—that is, the wrong decision by the machine to send a bag for further screening by the human screener and the lack of sufficient or appropriate information needed to resolve the alarm correctly using the TSA-established on-screen alarm-resolution protocol (OSARP). The TSA also indicated a desire to leverage what has been learned in the field of medical CT imaging and interpretation and to understand how this knowledge might apply to CT for explosives detection. In addition to the areas of focus described by the sponsor, the committee considered factors that may lead screeners to make a correct decision more often. The specific areas of focus addressed in the committee's review are identified in the section entitled "Screening Process" in Chapter 2.

Throughout the report the committee considers the trade-offs between false alarms and missed detections, but processing time, or throughout, is a third performance dimension. Indeed, with unlimited time to screen a bag, the estimations of mass and density would be near-perfect. The committee made the assumption that the amount of time currently required for bag screening is acceptable and focused on how to lower false alarms and maintain detection without explicitly paying a penalty in terms of much longer screening times. In implementing any changes intended to reduce false alarms, the TSA will have to determine if any increase in bag-screening time can be justified by a decrease in the number of false alarms.

---

[2] See, for example, Mathias S. Fleck and Stephen R. Mitroff, Rare targets are rarely missed in correctable search, Psychological Science 18:943-947, 2007; and Anina N. Rich, Melina A. Kunar, Michael J. Van Wert, Barbara Hidalgo-Sotelo, Todd S. Horowiths, and Jeremy M. Wolfe, Why do we miss rare targets? Exploring the boundaries of the low prevalence effect, Journal of Vision 8:1-17, 2008.

[3] Studies addressing the detection of novel and liquid explosives are discussed in the subsection entitled "Dual-Energy Scanning" in Chapter 2 of this report.

DISTINCTIONS IN TERMS AND THE NEED FOR DATA

Through discussions with the sponsor and site visits (see the section below entitled "Committee Meetings"), the committee learned that there are many mechanisms in the existing TSA system for recording data from EDSs—including the data on false alarms and their causes—but these data are not collected or analyzed in any comprehensive way. For this reason, although the causes of false alarms can be broadly identified, the committee is unable to assess the impact of any particular cause on either the overall false alarm rate or the system.[4] In particular, this lack of data also limits the committee's ability to assess what impact the false alarm rate has on personnel allocation or costs.

During the course of this study, the committee determined that there is inconsistent usage of the terms "false alarm rate" and "probability of false alarm." "False alarm rate" is a function of the number of bags scanned and the likelihood of non-threat items that can be confused with threat items. "Probability of false alarm" is a function of the specific contents of a scanned bag and the technology and decision algorithm of the EDS. The committee has tried to be consistent in using "false alarm rate" when comparing the number of false alarms to the total number of bags scanned, and "probability of false alarm" when discussing the chance that an individual bag will cause the EDS to alarm. Within the EDS community, the terms tend to be used interchangeably, but maintaining a distinction in the technical meanings of these terms is important in order to help the TSA take advantage of the large body of literature dealing with classification, detection, and statistical decision making.

The overall false alarm rate includes two distinct "populations" of bags, each of which would require a different approach in order to reduce false alarm rates:

- The first population includes bags for which the EDS cannot make a decision—so-called "exceptions," such as bags containing solid objects that cannot be penetrated by the EDS x-rays, mis-tracked bags, and bags that are poorly positioned in the EDS in such a way that the EDS cannot interrogate the entire bag ("cut bags"). These exceptions are sent directly to the bag inspection room (BIR) without the opportunity for a screener to evaluate the image and clear the bag.
- The second population includes bags whose contents include items that are misidentified by the EDS as potential threat items—for example, when the item's properties fall within the window defined for threat items, or multiple items are mistakenly agglomerated in the image and are treated as a single item that meets the criteria for a potential threat item. This second population is the one evaluated to determine the probability of false alarm.

The false alarm rate of the first population cannot be improved by improving the performance of the EDS because of limitations of the CT approach or of the baggage-handling system. The second population can be addressed by improving the ability of the EDS to interrogate the bag—for example, by improving the ability to distinguish threat materials from non-threat materials or by improving the ability of the image analysis algorithms to segment objects within a bag correctly. Although both of these populations contribute to the cost of evaluating false alarms, the committee focused on the second population, in which improvements to EDS technology can have an impact.

The committee heard estimates of the false alarm rate from many sources—the DHS, the TSA, airport operators, and equipment vendors. However, the committee was not able to find hard data for the entire system beyond estimates of the overall false alarm rate. Even such fundamental data as the percentage of bags scanned that were identified as "exceptions" were not readily available. In the absence of such hard data, the committee directed its focus primarily on machine-driven false alarms that lead to a bag's being sent to the BIR, as the additional time, personnel, and space necessary to resolve these alarms are likely a large component of the associated cost.

---

[4] For example, false alarms caused by hardware problems or errors in image reconstruction (see the subsection entitled "Image Reconstruction and Correction" in Chapter 2) cannot currently be distinguished from those caused by the automated threat-recognition algorithm (see the subsection entitled "Material Density" in Chapter 2).

STUDY PROCESS

Over the course of its study, the committee held four meetings. At its first meeting, on October 23 and 24, 2008, in Washington, D.C., the committee discussed the statement of task for the study and heard the perspectives of the TSA from its representatives as well as the perspectives of others regarding issues related to EDSs and false alarms.

The committee's second meeting was held February 11 to 13, 2009, in San Francisco, California. As part of this meeting, the committee visited San Francisco International Airport where it observed baggage-handling and baggage-screening operations from bag check-in through scanning, the OSARP and inspection processes, and, finally, the clearance of bags for loading onto an airplane. The committee spent the 2 days after the site visit in information gathering. On February 12, it heard from the DHS S&T on its view of the statement of task, and then from employees of L-3 Communications and General Electric (GE) Security, who spoke about their approaches to CT algorithm development and false alarm reduction.[5] On February 13, the committee heard from two experts in human-machine interaction in the medical CT field.

At its third meeting, held on April 27 and 28, 2009, in Washington, D.C., the committee heard from employees of the DHS and the TSA about the TSA's approach to deploying new technology. The committee then heard from the National Safe Skies Alliance[6] about its research into false alarms for carry-on baggage. Following this presentation, as part of its efforts to understand how researchers in the realm of EDS CT can learn from advances in medical CT, the committee heard from an employee of the U.S. Food and Drug Administration (FDA), who discussed the FDA's process of ongoing quality assurance for medical diagnostics. On the second day of this meeting, the committee heard from AAI Corporation (a Department of Defense contractor) regarding the use of performance-based incentive contracts. A speaker from Reveal Imaging then discussed that company's approach to false alarm reduction. After the meeting, two members of the committee participated on April 29, 2009, in a visit to the TSA Systems Integration Facility at Ronald Reagan Washington National Airport outside Washington, D.C.

The committee held its final meeting on June 15 and 16, 2009, in Woods Hole, Massachusetts, to draft the final report and reach a consensus on its conclusions and recommendation.

REPORT STRUCTURE

Following the report Summary—which includes key conclusions, findings, and recommendations from the chapters of the report—and the background provided in this introductory chapter, Chapter 2 presents an overview of CT-based EDSs and their integration in the airport setting. Chapter 3 discusses possible alternative approaches to false alarm reduction in the field: the use of multiple CT scans to improve the probability of detection, the use of mass spectrometry, x-ray diffraction technology, and the effective use of orthogonal technologies.

In Chapter 4, other approaches to complement improvements in technology—namely, possible adjustments to the contractual structure that the TSA uses with equipment manufacturers—are addressed with respect to decreasing false alarms in an airport setting.

---

[5] A representative from Reveal Imaging, another large vendor of CT-based EDSs, was unable to attend this meeting but did speak at the committee's third meeting.

[6] As described on its website (http://www.sskies.org, accessed December 3, 2009), "The National Safe Skies Alliance is a 501c 3 non-profit organization, formed in 1997 to support testing of aviation security technologies and processes." It is funded through the Federal Aviation Administration, and receives coordination support from the TSA. It conducts tests operationally at airports and pre-operationally at its facility in Alcoa, Tennessee.

In response to the sponsor's desire to capitalize on lessons from the medical CT realm, Chapter 5 provides an overview of medical CT systems and an analysis of approaches that may and may not be appropriate to incorporate into CT-based EDSs.

Finally, to provide the sponsor with guidance on how to make better use of the already-existing data in order to gain a better understanding of, and to be better able to address, the causes of false alarms, Chapter 6 focuses on issues related to data collection and analysis.

The appendixes of the report provide the following information. Appendix A presents biographical sketches of the members of the committee. Appendixes B, C, and D, by individual committee members, are independently authored papers with the endorsement of the rest of the committee. Appendix B outlines an approach to quantifying the risk and causes of false alarm scenarios associated with the airport screening of checked baggage. Appendix C discusses chemistry-based alternatives to computed tomography-based explosives detection. Appendix D presents a statistical approach to reducing the probability of false alarms while improving the probability of detection. Appendix E presents the committee's statement of task. Finally, Appendix F provides a list of the acronyms used in the report and their definitions, as well as a brief glossary.

2

Overview of Deployed Explosive Detection System Technologies

This chapter discusses computed tomography (CT) technology as it is applied to explosives detection (image reconstruction and the automated threat recognition [ATR] algorithm), ways in which the physics of CT limit its applicability for explosives detection (including image artifacts and material density), the processes of testing and evaluating these systems at the Transportation Security Laboratory (TSL), and the implementation of explosive detection systems (EDSs), including the screening process, in an airport setting. The differences between CT as used for explosives detection and medical CT are covered in Chapter 5 of this report.

As pointed out in Chapter 1, the equipment described in this report does not detect the presence of explosives. Instead, as used by the Transportation Security Administration (TSA) and this committee, the term "explosive detection system" refers to a CT-based device for interrogating a bag, coupled with an ATR algorithm for evaluating the results of the interrogation. The purpose of the algorithm is to identify materials within checked bags that possess certain known properties of explosives, in order to determine whether or not an alarm should be given. All EDSs must be certified by the TSL, as discussed in the subsection below entitled "Testing at the Transportation Security Laboratory."

OVERVIEW OF A COMPUTED TOMOGRAPHY SCANNER

A typical CT scanner (Figure 2-1) consists of a support frame and five key subsystems: (1) a high-voltage power supply (HVPS), (2) an x-ray tube, (3) a detector, (4) a gantry, and (5) a data acquisition system (DAS). Bags are fed into the scanner by means of a conveyer belt.

The HVPS produces voltages necessary to power the x-ray tube. The potential of the HVPS generally falls between 140 and 180 kilovolts, with power in the range of 500 to 5,000 watts. Some systems use a direct current waveform. Other systems add an alternating current component as one means of collecting dual-energy information (see the subsection below entitled "Dual-Energy Scanning").

The x-ray tube produces a Bremsstrahlung spectrum of x-rays from 0 kiloelectronvolts to the peak potential of the HVPS.

One or more rows of detectors, outwardly aligned in a cone shape from the x-ray tube, first convert the x-ray photons into light photons, and then convert the light photons to an electrical charge. The output of the detector is then digitized by the DAS into either fan- or cone-beam projections. These projections are related to the line-integrals of the x-ray attenuation coefficient of the bag along the paths from the x-ray tube to the detectors and are sampled at approximately 1 kilohertz so that projections are obtained at various angular positions around the bag.

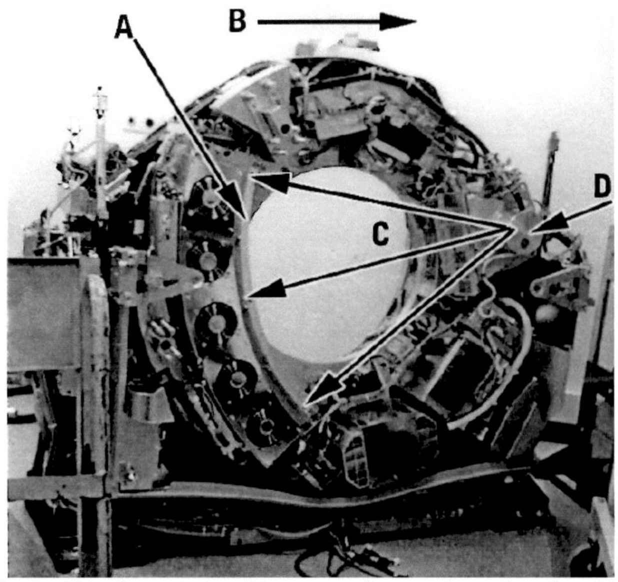

FIGURE 2-1 Photograph of the inside of a computed tomography scanner, showing (A) x-ray detectors, (B) gantry rotation, (C) x-ray beam, and (D) x-ray tube. The data acquisition system is not shown. SOURCE: Adapted from a Wikimedia Commons image available at http://en.wikipedia.org/wiki/File:Ct-internals.jpg.

The x-ray tube, power supply, detectors, and DAS are mounted on a gantry. The rotation speed of the gantry ranges from 60 to 180 revolutions per minute.[1] As the gantry rotates around the bag, the conveyor belt on which the bag rests may be stationary or moving. If the conveyer is stationary during scanning, the scanner is considered to be a "step-and-shoot" variety. Conversely, if the conveyer is moving, the scanner is helical or spiral.

Image Reconstruction and Correction

Following the scan, the outputs of the DAS are sent to a reconstruction computer (see Figure 2-2) to be converted into cross-sectional images, which are then sent to another computer on which the automated threat-recognition algorithm is performed.

Most scanners use a process called filtered back-projection (FBP) to reconstruct the cross-sectional images;[2] the algorithms used in image reconstruction were developed for medical imaging and have not been optimized for use in security applications, which is one potential source of error that can lead to false alarms.[3] In the reconstruction process, the output of the DAS is corrected to account for imperfections in the machine hardware and to generate the line-integral data. Additional steps are used to compensate for the cone shape of the x-ray beam and for helical scanning (where the scanner, itself, moves). These steps are approximate, and artifacts are created in images due to these approximations. The steps in the reconstruction process are shown in Figure 2-2.

---

[1] Or 0.33-1.0 second per rotation. In general, newer models are faster. See, for example, Ge Wang and Hengyong Yu, An outlook on x-ray CT research and development, Medical Physics 35(3):1051-1064, 2008.

[2] A.C. Kak and M. Slaney, Principles of Computerized Tomographic Imaging, IEEE Press, New York, N.Y., 1988.

[3] See, for example, Lawrence Livermore National Laboratory, Improved Aviation Security via Technology Advancements, available at http://www-eng.llnl.gov/ndc_aviation.html, accessed April 7, 2011, which describes the shortcomings of these medical CT-based reconstruction algorithms and the work being done at Lawrence Livermore National Laboratory to improve them.

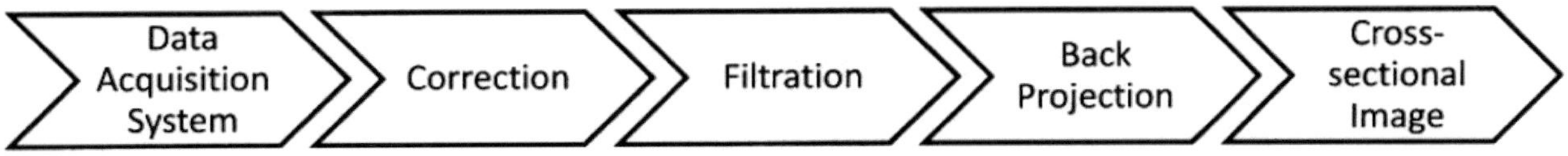

FIGURE 2-2 Steps in the computed tomography image reconstruction process.

TABLE 2-1 Operations Performed During the Correction Steps in Image Reconstruction

Step	Synopsis
Offset	The electronics (photodiode and amplifiers in the data acquisition system [DAS]) have dark currents. The dark currents are measured with the x-ray tube turned off and then subtracted. Temperature drift of the offset has to be considered.
Reference	The current supplied by the high-voltage power supply to the x-ray tube may vary. A reference detector measures the incident x-ray flux.
Beam hardening	The x-ray tube produces a polychromatic spectrum. The x-ray attenuation coefficient is a function of the photon energy, with lower-energy photons being preferentially removed. A polynomial correction is applied. Unfortunately the different materials require the use of different polynomials, and so artifacts will remain.
Spectral response	Each detector has its own spectral response to polychromatic x-rays. This response is known as the detector's transfer function. The difference of the transfer function for each detector with respect to the mean of the functions for all the detectors is corrected in order to prevent the insertion of concentric rings and bands in images. One method of performing this correction is to apply a polynomial correction, which is specific to each detector and whose coefficients are determined during a calibration step.
Afterglow	The detector and DAS have finite impulse responses leading to a temporal blur of the projections. The impulse responses may be de-convolved.
Scatter removal	Scattered x-ray photons may reach the detector. Some scattered photons may be eliminated with antiscatter plates placed in the septa between detectors. Additional algorithmic correction can be used to remove scatter based on measurements from auxiliary detectors or using the projections themselves.
Clamping	The DAS has a finite dynamic range, which is determined in part by the electronic noise in the DAS. The number of x-ray photons reaching the detector may be on the level of the electronic noise. The number of photons is clamped to a positive number. However, artifacts will still be generated in images when this condition occurs.
Gain measurement	Each detector has its own gain. The gain is measured by scanning only air. The values of the air readings are used to scale the readings through a bag. The gains may be a function of the angular position of the gantry.
Logarithm	The DAS/detector combination integrates energy. In order to generate the line integrals required by filtered back-projection, the natural logarithm of the reading is taken. Note that this step is based on the physics of detecting x-rays and is not a correction step. However, it is included in the correction steps of reconstruction because of implementation considerations.
Re-binning	The cone-beam projections are processed to form fan-beam or parallel-beam projections. If the projections were acquired using helical scanning, then the movement of the bag during data acquisition is removed using interpolation.

Imprecision in image reconstruction is one source of error that can lead to false alarms (see the subsection below on "Image Artifacts"). If the steps in correction cannot completely account for the underlying physical effects of scanning, images will be degraded. Another source of artifacts is due to the approximations made in the reconstruction algorithm for compensating for cone-beam divergence of the x-ray beam and for helical scanning. Any artifacts can lead to inaccurate measurements of an object's linear attenuation coefficient, density, and atomic number. Table 2-1 provides an overview of the operations that are performed during correction.

Noise in the image may also create problems in the automated image segmentation process, and make it difficult to distinguish between explosives and other material. While filters, image smoothing, and other enhancement techniques have been developed for medical CT applications, these measures can be difficult to extend to baggage, because of the large number of objects present in baggage and because—unlike tumors—a single object in a baggage scan may present with wide number of grey levels in the image.[4]

Automated Threat Recognition

The automated threat-recognition process (outlined in Figure 2-3) segments the cross-sectional images into individual objects and then classifies each object as either a threat or a non-threat. Specifics of each vendor's ATR algorithm are proprietary; the process described here is general. As part of its analysis, the ATR algorithm may also compensate for imperfect correction in the CT reconstruction step and extracts features such as density, atomic number, and feature size.

Once this information has been extracted, it is compared to the density and properties of known explosives. If the information derived from an object falls within the specified range and the object's mass is above a TSA-specified value for that range, then the object is declared a potential threat.

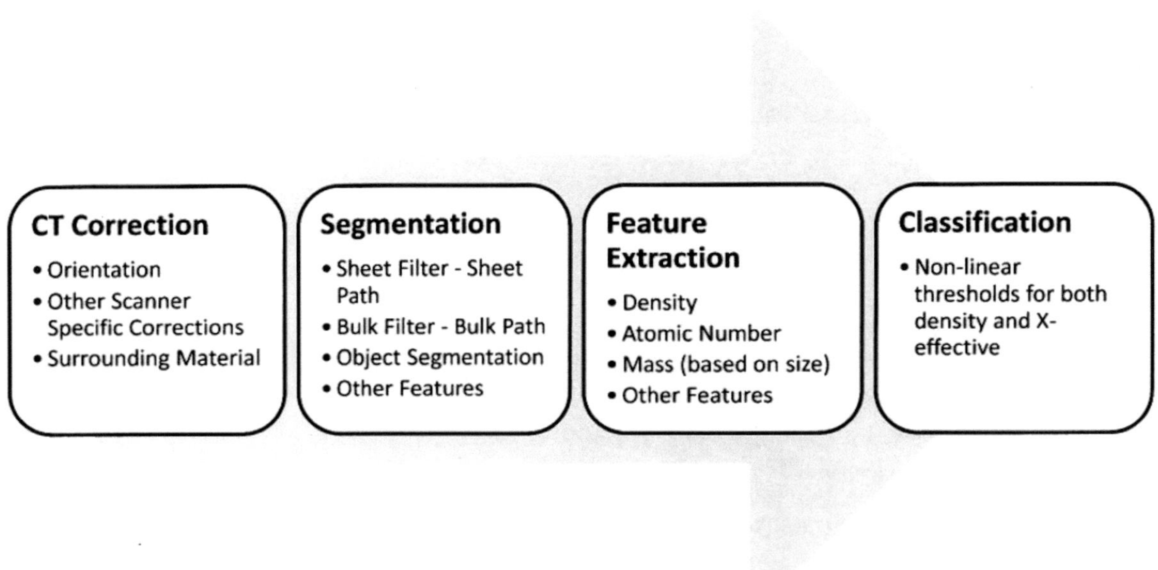

FIGURE 2-3 Simplified, representational outline of the automated threat-recognition process. Note that each explosive detection system vendor's process is different.

[4] Sameer Singh and Maneesha Singh, Explosives detection systems for aviation security: A review, Signal Processing 83(1):31-55, 2003.

	Alarm	No Alarm
Threat Present	True Detection	Missed Detection
No Threat Present	False Positive	True Negative

FIGURE 2-4 A contingency table of the potential results of an interrogation of checked baggage by a computed tomography-based explosive detection system.

Some vendors may also use the filtered back-projection data and the images from the digital radiography line scanner in their ATR algorithm. They may also have separate methods (or "paths") for the identification of potential sheet explosives and the identification of potential bulk explosives.

An alarm may be either a "true positive detection" (meaning that the ATR algorithm signals an alarm and a threat is present in the scanned bag) or a "false alarm" (sometimes called a "false positive," which means that the ATR algorithm signals an alarm but no threat is present). Bags that do not cause the EDS to alarm may be "true negative" (the ATR algorithm does not signal an alarm, and there is no threat present) or a "missed detection" (meaning that the ATR algorithm failed to report the presence of a threat). These possibilities are detailed in Figure 2-4.

The ATR algorithms have been developed and refined over many years to alert on threat amounts of materials that fall within a specified density and mass range ("detection window"). However, owing to the nature and composition of many non-threat objects, the criteria cannot be made specific enough to include *only* threat materials, and innocuous materials may fall within or near the detection windows and may be mistaken for threat materials. Consequently, there will always a trade-off between false alarms and missed detections.

When applied only to the information available from the CT scan, no algorithm will always identify a threat item while never misidentifying an innocuous item as a threat. Narrowing the detection windows in order to eliminate the misidentification of non-threat materials carries with it the risk of decreasing the detection rate and missing a true threat. Expanding the detection window to ensure the capture of all threat materials will result in capturing non-threat materials and increasing the false alarm rate. A plot of threat identification (probability of detection, or PD) versus misidentification of an innocuous item as a threat (probability of false alarm, or PFA) is known as the receiver operating characteristic (ROC) curve, as shown in Figure 2-5. A method of reducing the false alarm rate is to increase the area under the ROC curve; this can be done for the EDS by adjusting the signal-to-noise ratio (either increasing the strength of the signal or reducing the amount of noise), but the maximum area under the curve is fundamentally limited by the technology.[5]

[5] David Heeger, "Signal Detection Theory." New York University, 1997, available at http://www.cns.nyu.edu/~david/handouts/sdt-advanced.pdf, accessed June 14, 2011.

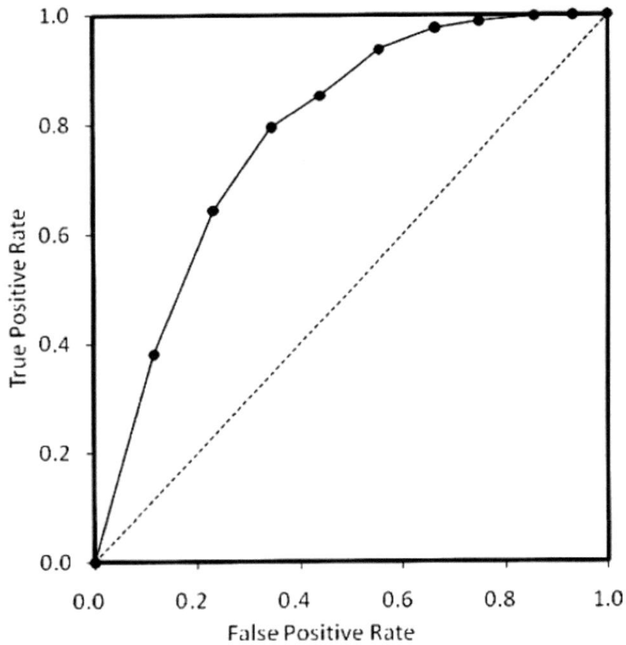

FIGURE 2-5 An example of a receiver operating characteristic curve.

The Transportation Security Administration has chosen to maintain a high level of detection. The agency thus has been forced to accept a concomitantly high level of false alarms.

Vendors who spoke at the committee's meetings (see the section entitled "Study Process" in Chapter 1) indicated that more recent hardware and software updates have the potential to deliver a lower false alarm rate while maintaining probability of detection, but they indicated that in many cases these technical improvements are not followed up on by the TSA or the airports.[6] For example, the committee was told that General Electric (GE) Security's CTX 9400 model[7] demonstrated a 45 percent reduction in "shield alarms" (that is, alarms that occur when the CT machine cannot penetrate an area of a bag) and a 10 percent reduction in other false alarms while concurrently demonstrating better detection and a slightly lower throughput (fewer bags per hour).[8] However, the cost to upgrade each of these machines is more than $100,000, which has—up to this point—been regarded as prohibitively costly for the TSA or airports. L-3 Communications also indicated that it had developed TSL-certified software to reduce false alarms, but that, as of the committee's meeting in 2009, this software had not been purchased.

FUNDAMENTAL LIMITATIONS OF COMPUTED TOMOGRAPHY-BASED EXPLOSIVE DETECTION SYSTEMS BASED ON THE PHYSICS OF THE TECHNOLOGY

As stated above, CT-based explosive detection systems are not systems that detect explosives, but rather systems that can identify materials that have specific properties. In this section the committee expands on the limitations of the use of this form CT for the purposes of detecting explosives.

---

[6] David Heeger, "Signal Detection Theory." New York University, 1997, available at http://www.cns.nyu.edu/~david/handouts/sdt-advanced.pdf, accessed June 14, 2011.

[7] Now produced by Morpho Detection, Incorporated.

[8] Matthew Merzbacher, General Electric, "Overview of Detection Algorithms," presentation to the committee. February 12, 2009, San Francisco, Calif.

Image Artifacts

The correction step in CT image reconstruction attempts to compensate for imperfections in the projection data acquired during scanning, but these corrections are not perfect, and image artifacts will always remain. These artifacts produce signals that can lead to uncertainty in the measurement and evaluation of the objects that are being scanned. This uncertainty manifests itself in the threat-detection process in various ways—for example, in mis-estimation of object mass, a widening of density and atomic number windows, and inaccurate region building (which leads to over-aggregation of different objects). This issue, which is also relevant to medical CT, is discussed in Chapter 4.

Image artifacts contribute to false alarms mainly by causing uncertainty in the shapes and sizes of the individual objects within a bag. Although metal—which leads to streak artifacts in CT images—is one of the primary sources of image artifacts, there are other contributing factors, such as photon starvation (common in large, densely packed bags or bags that include one or more heavy metal objects), beam hardening (generally caused by long, very straight objects that run across the width of a bag), partial volume (caused by very thin objects), bag motion artifacts, and approximations in the reconstruction algorithm to compensate for cone-beam divergence and helical scanning. The larger and more cluttered a bag is, the more likely there are to be errors in image reconstruction.

Image artifacts caused by imperfections in both the CT hardware and the software reconstruction lower confidence in the estimated characteristics of an object within a bag, forcing the threat-defining windows to be widened, which results in a concurrent increase in false alarms. Thus, improvements in the image reconstruction and correction process would enable the more accurate measurement of objects and could lead to a lower false alarm rate.

Baggage Contents

Variations in the attitudes and practices of the flying public can also lead to variations in the false alarm rate over time. EDS vendors and screeners at San Francisco International Airport told the committee that different airline policies and TSA policies have had an impact on the way that passengers pack their bags and, consequently, on the false alarm rate. For example, when more airlines began to charge for checked baggage customers responded by packing their suitcases more densely, which, as noted above, can also lead to a higher false alarm rate.

Seasonal changes can also precipitate changes in passengers' packing habits and can make it difficult to know where to set the parameters for the ATR algorithm. For example, passengers traveling in the summer or to a tropical destination and those traveling in the winter or to a snowy destination are likely to pack differently.

Material Density

A non-threat material that falls within the same density and atomic number range as that of a given threat material will result in a false alarm. The difficulty in isolating threat materials from non-threat materials can be seen in Figure 2-6, which shows typical density ranges for some threat and non-threat materials. This figure shows that while clothes are clearly distinguishable as a single-valued function of density, other non-threat materials commonly found in passenger bags overlap in material density with some threat materials. It is inevitable then that relying solely on density to indentify threat material will lead to some misidentification and false alarms.

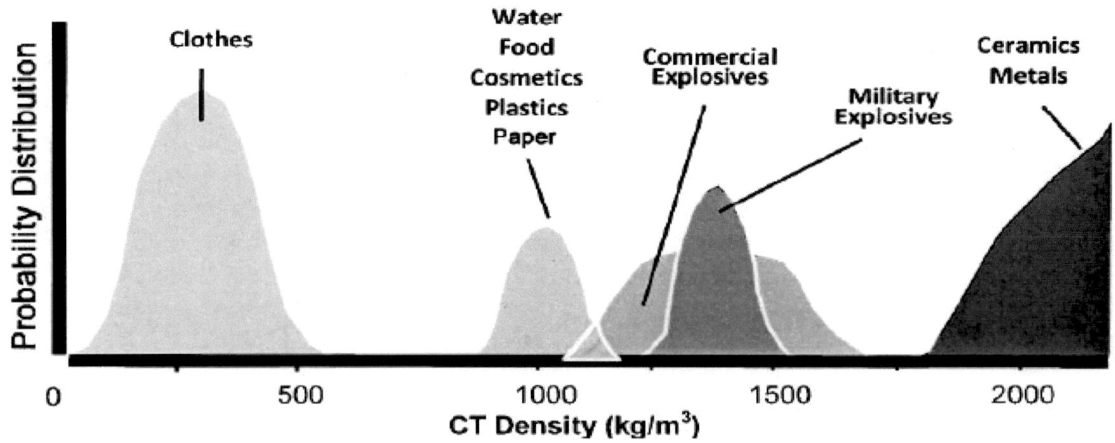

FIGURE 2-6 Notional distribution of threats and non-threats in computed tomography (CT) density space. Clothes are clearly distinguishable as single-valued function of density, but other non-threat materials commonly found in passenger bags show some overlap in material density with some material density with some threat materials.

Setting the limits for declaring a material a potential threat is a difficult balance of science and policy. For example, a lower density limit of 1,100 kilograms per cubic meter would lead to missing the detection of some commercial explosives, whereas many innocuous materials would still be identified as potential threats. Lowering the limit to about 1,000 kilograms per cubic meter would allow capturing more commercial explosives, but it would also result in misidentifying a much larger number of innocuous materials. Adding the atomic number may reduce the overlap in two-dimensional space by providing conditional data to distinguish between threat and non-threat materials, as discussed in the next subsection.

Dual-Energy Scanning

Because errors in indentifying non-threat materials as potential threat materials occur when the non-threat materials have a density similar to that of threat materials, adding atomic number to the screening criteria could improve the ATR algorithm's ability to distinguish between threat and non-threat materials and lower the probability that the EDS would give a false alarm. Although it seems reasonable that adding an extra dimension to the measurement would improve false alarm rates, anecdotal information presented to the committee indicates that the false alarm rates for dual-energy CT machines in an airport setting are not appreciably different from the false alarm rates for single-energy machines.

Nevertheless, the committee does believe that the technology deserves further exploration so that there can be a full understanding of its advantages and limitations. Reveal Imaging has conducted a DHS sponsored study to gain a better understanding of limits of its CT-80 machine and false alarm images from two different airports.[9] This is a very important first step to allowing researchers and the TSA to evaluate more effectively the potential improvements and results of their efforts.

TESTING AT THE TRANSPORTATION SECURITY LABORATORY

Certification testing of EDSs and their subsequent performance testing in an airport setting are one way to gain a better understanding of EDS performance and of the causes of false alarms. To be

[9] Elan Scheinman, Reveal Imaging Technologies, Inc., presentation to the committee, April 29, 2009, Washington, D.C.

certified, a machine must demonstrate the ability to detect a number of categories of explosives, with each category having a specific detection threshold (i.e., level of detection that must be met.) The machine must also meet an average detection threshold across all categories of explosives and not exceed a maximum false alarm rate (which is tested separately from the detection).

This certification testing is performed at the Transportation Security Laboratory (TSL), located at the William J. Hughes Technical Center at the Atlantic City International Airport in New Jersey. Originally, established in 1992 as part of the U.S. Department of Transportation, the laboratory is now under the umbrella of the U.S. Department of Homeland Security Science and Technology Directorate.

There is a limitation in being able to predict the performance of machines in an airport setting; however, the use of bags more representative of those seen in an airport setting would be enormously complicated, given the variations in bags and contents. Additionally, the use of a specific set of test bags allows all manufactures to be tested against a common standard.

Other tests besides that for certification are performed at the TSL. Certification readiness testing and pre-certification tests qualify a system to enter the path to certification. Post-certification tests are performed to ascertain whether certain configurations or locations of explosives go undetected.

Conclusion: Certification testing at the Transportation Security Laboratory fills a specific and useful role. However, it should not be used as the sole basis for predictions of performance in an airport setting.

IMPLEMENTATION WITHIN AN AIRPORT SETTING

Shifting Emphasis

In the airport setting a shift in emphasis occurs—from a focus on detection to a focus on reducing the costs of screening by minimizing the number of secondary inspections and the number of manual bag inspections. Because on-screen alarm resolution is the link between the automated alarm of the EDS and manual bag inspection, there can be pressure to clear bags in order to make delivery deadlines.[10] Additionally, because the bags and objects scanned in the airport are more varied than those in the TSL-certified test set, the probability of detection established for an EDS at the TSL may not be maintained an airport setting. Beyond these limitations, additional shortcomings in the effectiveness of an EDS system may develop over time, including, for example, the lack of a feedback system by which false alarms are analyzed and fed back into the ATR software development (discussed in greater detail in Chapter 6).

Recommendation: The TSA should develop procedures for periodic verification to ensure that fielded EDSs meet detection-performance-level standards that correspond to the requirements for EDS certification. In addition to monitoring detection capability directly (e.g., using standard bag sets and red-team testing), these procedures should include the frequent monitoring of critical system parameters (e.g., voltages and currents) and imaging parameters (e.g., image resolution and image noise) to detect system problems as soon as they arise. For purposes of monitoring EDS performance, the TSA and EDS vendors should develop specification limits for all critical system parameters (and their tolerances) that could be monitored frequently and recorded to track changes in performance during normal operations or to verify performance after maintenance or upgrading.

---

[10] See, for example, Sara Kraemer, Pascale Carayon, and Thomas F. Sanquist, Human and organizational factors in security screening and inspection systems: Conceptual framework and key research needs, Cognition, Technology, and Work 11:29-41, 2009.

Screening Process

Airports can incorporate explosive detection systems into their screening processes in a variety of ways, depending on such factors as the space available, flight schedules, and typical flight destinations. Here the committee describes a "typical" screening scenario as an aid to understanding the overall checked-baggage screening system, including alarm resolution by human screeners. This scenario is not meant to represent a preferred approach to screening or to limit the variation in checked-baggage screening equipment in airports.

EDSs are deployed in two basic configurations: (1) in-line: the bags are fed into an EDS by the baggage-handling system,[11] and (2) stand-alone: the bags are fed into an EDS manually. The stand-alone EDSs are usually in the check-in lobby or behind the check-in counter.

Figure 2-7 represents the CT-based EDS screening process. It should be noted that most of the expenses associated with clearing false alarms occur at the baggage-viewing station and in the baggage-inspection room (BIR). Clearing alarms in both of these areas requires human intervention in the process.

When the ATR algorithm determines that an object or objects within a scanned bag meet the established threat criteria, a human screener must resolve the alarm. Information is presented on a display to a human screener at a baggage-viewing station. The information includes cross-sectional images of the bag and specifies of any suspicious objects generated by the ATR algorithm. The screener uses this information to decide either (1) that the alarm was caused by a non-threat item (whereupon the bag can be cleared to go on the airplane) or (2) that the screener is unable to determine—based on the on-screen alarm-resolution protocol (OSARP)—that the object indentified by the machine is not a threat, in which case the bag is sent to the baggage inspection room or other local area where it is opened and examined manually.

As part of its analysis, the ATR algorithm evaluates whether there are any areas of the bag that the x-rays cannot penetrate. If there are any such areas, the system signals a shield alarm. Because these "shielded" areas could conceal potential threats, any bag with a shield alarm is sent directly to the baggage-inspection room for further screening, bypassing the option of on-screen resolution. In addition to shield alarms, "exceptions" that result in a bag's being sent directly for further screening include "mis-tracking" (the baggage-handling system loses track of the bag), operator time-out errors (the operator fails to clear the bag within a time limit), jamming of the bag in scanner, and scanner failures. In discussions with the committee, screeners indicated that these exceptions are counted as part of the overall false alarm rate, but specific data related to the percentage of the overall rate that they represent were not available.

The on-screen alarm-resolution protocol serves as the link between the ATR algorithm and the baggage-inspection room.

Adjusting the operating point on the ROC curve of the ATR algorithm and introducing variability in the EDS's performance may improve the overall performance of the system, which includes the decision by both the EDS and the screener. This variability could be driven by intelligence-adjusting the algorithm to be more sensitive toward specific types of explosives while not searching for those that are less likely to be used. Or it could be driven by the introduction of other data to create a passenger-specific, risk-based screening approach. Such variability might also provide might also provide a deterrence value, as it would make it more difficult for any adversary of the system to predict the EDS's capabilities.

It will remain necessary to include the screener-in-the-loop when making any modifications to the overall screening system's operations. Any changes may also require changes to the on-screen alarm-resolution protocol to ensure that the link between the ATR algorithm and the baggage-inspection room functions effectively. At risk assessment approach, such as the quantitative risk assessment process

[11] The baggage-handling system consists of a set of conveyor belts, diverting mechanisms, and a tracking system. The conveyor belt moves bags in and out of the EDS, to the baggage-inspection room or other local area, and to the airplane. Diverting mechanisms transfer the bags between the different sections of the conveyor belts.

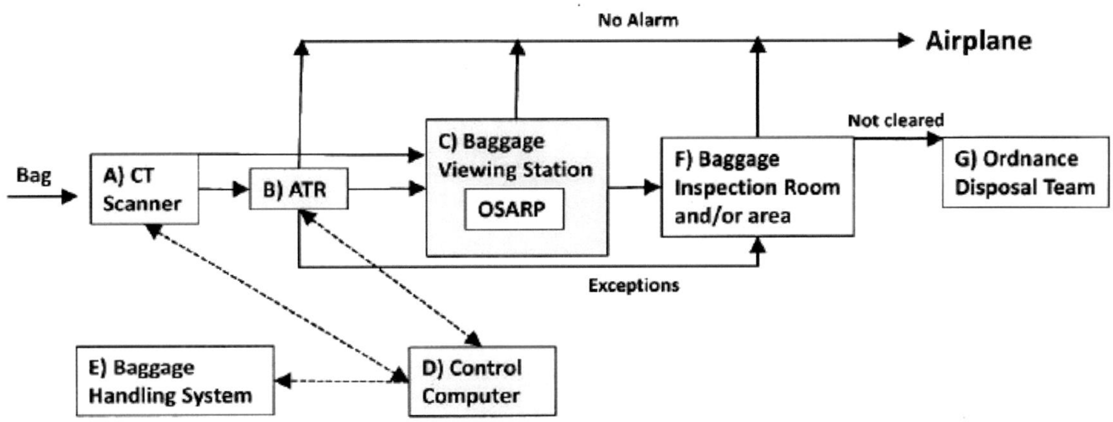

FIGURE 2-7 Diagram of an in-line explosive detection system (EDS) consisting of (A) the computed tomography (CT) scanner, (B) the automated threat recognition (ATR) algorithm, (C) the baggage-viewing station and the on-screen alarm-resolution protocol (OSARP), and (D) the control computer. The EDS is integrated with (E) the baggage handling system, (F) the baggage-inspection room and/ or area, and (G) the ordinance disposal team. Shaded boxes are components of the EDS, white boxes are subsystems used in conjunction with the EDS, solid connecting lines show the flow of bags and/ or images of the bags, and dashed connecting lines show the flow of the control and information.

described in Chapter 6 and Appendix B in this report, may be one way to evaluate these changes. Because the probability of false alarm can be measured in an airport setting, and probability of detection is rarely measured except in Transportation Security Laboratory testing, it is difficult to determine the simultaneous PD/PFA performance of EDSs in an airport setting. The committee believes that it is likely that the PD measured at the TSL is not maintained in an airport setting owing to a combination of the use of non-representative bags to measure PD at the TSL and the possibility that screeners may clear too many bags on screen that should be inspected by hand due to the low occurrence of true positives. Paradoxically, forcing EDSs to operate at their highest PD, and simultaneously their highest PFA, creates a situation in which screeners expect that every alarm is a false alarm and in which bags are cleared that should be sent for additional screening, lowering the detection capability of the entire system. It is counterintuitive that lowering the probability of detection of the EDS could lead to increased probability of detection of the overall system, but the committee believes that the TSA should consider evaluating this possibility.

When a bag is sent to the baggage-inspection room, screeners may open it to visually inspect the objects indentified as potential threats and—depending on the object indentified—may also employ explosive trace detection to attempt to clear the bag. If the airport's integration of the baggage-inspection room with the rest of the baggage-screening system is robust enough to permit it, this inspection may be guided and informed by other data related to the bags being inspected including CT slices and the outputs of the ATR algorithm-although it is possible that the threat indentified by the ATR algorithm will not be found by the transportation security officer (TSO) screening the bag or that another item will be mistaken for the threat. If the screener is able to clear all potential threats in the bag, it is sent to the airplane. Bags that cannot be cleared are handled according to local regulations for potential explosive threats.

Finding: The low prevalence of true positives may make it nearly impossible to measure probability of detection with humans-in-the-loop without forcing true positives via red-team testing.

Recommendation: The Transportation Security Administration, through the Transportation Security Laboratory, should support human-factor studies to assess the impact on overall system performance, that is, the EDS plus the screener resolution, when the operating point on the explosive detection system's receiver operating characteristic curve is adjusted so that both the probability of detection and probability of false alarm are lowered. If the results of such studies determine that screener attention is degraded by the expectation that every alarm is a false alarm, the TSA should consider implementing adjustments to the operating point on the receiver operating characteristic curve and allowing vendors to reduce probability of detection in an airport setting to the minimum rate required for certification.

DISCUSSION, WITH RELATED FINDING AND RECOMMENDATION

To reduce the costs associated with screening baggage in airports, it will be necessary to lower both the number of automated alarms from EDSs and the number of bags opened manually and the number of bags that have to be traced. The automated threat-recognition algorithm will be able to make more correct decisions when it is provided with more accurate information about the contents of bags, including both the materials properties and object sizes. Improving object segmentation so that adjacent but unrelated objects in the bag are correctly separated in the image could also reduce false alarm rates by allowing for a more accurate estimation of an object's density and mass. Hardware improvements aimed at more accurate estimates of materials properties could include dual- or multi-energy approaches or other methods to differentiate materials within a bag.

Modifying an EDS's operation to emphasize image quality over operational requirements (such as throughput) by such means as slowing the scan speed, improving reconstruction algorithms, or changing parameter settings within ATR algorithms based on threat level could also achieve the same aim—that is, it could improve the estimate of materials properties or object segmentation. Augmenting the CT scanner data with additional screening data could provide additional means of distinguishing between threat and non-threat materials.[12]

Results of inspections in the baggage inspection room would be useful in identifying the causes of false alarms, but in the committee's tour of the BIR at San Francisco International airport the committee saw no mechanism for collecting data on the results of such inspections. The committee believes that San Francisco International Airport is representative of airports through the Unites States in this respect. However, as with the data on exceptions, there is no requirement of either the TSA or the EDS vendors to mine or analyze collected data. This topic is discussed more fully in Chapter 3.

Finding: Based on the information available at this time about the performance characteristics of these approaches and available data on the actual sources of false alarms raised by today's explosives detection systems, it is not possible to establish which are most promising or merit significant investment.

Recommendation: The TSA should not fund an overall replacement of fielded explosive detection systems, because replacing all the units in service with currently available technology would not allow for learning in an airport setting to inform future performance improvements. Instead, the TSA should plan its capital spending for explosives detection improvements over a period of time sufficient to allow several generations of technology to be fielded on a limited basis, revaluated, and iteratively improved—thus leading to a gradual improvement in the overall field performance of CT-based explosives detection systems.

---

[12] A full discussion of this process can be found in National Research Council, Fusion of Security System Data to Improve Airport Security, The National Academies Press, Washington, D.C., 2007.

3

Alternative Approaches for the Reduction of False Alarms

In this chapter the committee encourages the Transportation Security Administration (TSA) to look beyond CT in addition to driving computed tomography (CT) to its best performance and improving the algorithms for detecting threat items in the resulting images. The chapter discusses four potential approaches to reducing false alarms in the field: (1) using multiple CT scans to improve the probability of detection (PD), (2) using mass spectrometry, (3) employing x-ray diffraction technology, and (4) and incorporating data from other sources. The material presented is preliminary, but further consideration and study of these approaches have the potential to positively affect the efforts to reduce false alarms in the screening of checked baggage in U.S. airports.

AN ALTERNATIVE APPROACH: MULTIPLE SCANS USING EXISTING TECHNOLOGY

One of the ways to improve the performance—that is, to reduce the probability of false alarms (PFAs) and to improve the probability of detection—of CT-based explosive detection system (EDS) scans is to increase the number of cross-sections that the machine takes of an object. More cross-sections usually lead to a better probability of correct discrimination in recognizing whether an object is a threat or a non-threat. The number of cross-sections that a machine takes can be increased either by changing the current hardware or by passing a bag through the CT scanner multiple times in such a manner that the bag is positioned somewhat differently for each scan.

When a bag is passed through the CT scanner multiple times, a natural decision-making situation arises—that is, a decision rule is needed with respect to when a bag should be declared a possible threat and sent for manual inspection. The committee offers the following statistical model to present the idea that false alarms can be reduced using the current hardware in a different way. Although according to this model the same bag is being scanned multiple times, the presumption is that neither the machine nor the operator knows that it is the same bag, and in this way the scans remain "independent." This is reasonable because the bag will most likely be positioned somewhat differently at each pass because of the bumps on the conveyor belt. This may not be possible for unusually large items; however, many of them are manually processed already. This "independence" can also be increased by using multiple machines for the different scans each of which are set to provide cross-sections at slightly different depths.

Suppose that a bag is scanned N number of times and out of these N scans, the machine alarmed on it as a potential threat q number of times. The natural questions that arise at this stage are as follows: (1) How many times should a bag be scanned (that is, what is the value of N)? and (2) At what value of q should the bag be declared a potential threat that should be sent for manual inspection?

Now suppose that a bag, in fact, contains a threat object that can potentially be detected by a CT scanner. Suppose that the probability of detecting the threat object in a single scan is δ. In this hypothetical, the total number of scans is fixed as N and the decision rule is that the bag is declared a threat (and hence should be sent for manual inspection) if the bag is alarmed q times out of N scans. Then assuming that the scans are independent of each other, this can be computed by using the binomial distribution[1] as

$$(\mathrm{U}) \quad \sum_{i=q}^{N} \binom{N}{i} \delta^{i}(1-\delta)^{N-i} = P_D \,.$$

This is the probability of correct detection under the decision rule specified above.

Similarly, suppose that the bag does not contain a threat object, but a single CT scan may declare it falsely as a potential threat with probability α. Then, the probability that such a bag will be sent for manual inspection is again obtained by using the binomial distribution as

$$(\mathrm{U}) \quad \sum_{i=q}^{N} \binom{N}{i} \alpha^{i}(1-\alpha)^{N-i} = P_F \,.$$

This is the probability of false alarm under the decision rule specified above.

Different values of *N* and *q* lead to different PDs and PFAs. The detailed description of the solution and computational details are provided in Appendix D in this report. Table 3-1 presents the probability of false alarms and probability of correct detection under different values of *N* and *q* in the above decision rule.

TABLE 3-1 Probabilities of Correct Detection and of False Alarm for Some Combinations of *N* and *q*

N	*q*	Correct Detection	False Alarm
2	1	0.99	0.36
	2	0.81	0.04
3	1	0.999	0.488
	2	0.972	0.104
	3	0.729	0.008
4	1	0.9999	0.5904
	2	0.9963	0.1808
	3	0.9477	0.0272
	4	0.6561	0.0016
5	1	0.99999	0.67232
	2	0.99954	0.26272
	3	0.99144	0.05792
	4	0.91854	0.00672
	5	0.59049	0.00032

NOTE: These probabilities assume that each scan represents an independent interrogation of a random object. In this scenario, $\alpha = 0.2$ and $\delta = 0.9$.

[1] G. Casella and R.L. Berger, Statistical Inference (2nd ed.), Duxbury Press, Pacific Grove, Calif., 2002.

A typical entry in the table is read as follows: Suppose that the decision rule is such that a bag is declared a threat if it tests positive at least three times out of the total of five scans. Then such a decision rule will detect the threat correctly 99.14 percent of the times and will give a false-positive alarm 5.79 percent of the times. Presuming the committee's assumption regarding the independence of the scans is correct, it is clear that multiple scanning can reduce the probability of false alarms, at the same time increasing the probability of correct detection of threats (those that can potentially be detected by a CT scan).

If scans are automated it will allow for a greater likelihood of accurate results with no additional personnel costs. However, there will be other costs associated with automating the rescanning, such as increased screening time and additional routing hardware, as well as the costs of tracking the multiple scans of the same bag, and these must be offset by an improvement in false-alarm rates.

ANOTHER ALTERNATIVE APPROACH: CHEMICAL ANALYSES

One task in the charge to the committee was an examination of the problem of explosives detection by new non-certified methods. This section deals with the use of mass spectrometry to address this charge. Box 3-1 features an extract a 2004 National Research Council report on the subject.

Research programs, both internal to TSL and externally supported, have examined the capabilities and potential screening application of mass spectrometers. Nevertheless, there exists the perception that mass spectrometers are too complex, difficult to operate, and insufficiently rugged for deployment. Recent advances in mass spectrometry—mostly made since 2004—have dramatically changed the capabilities of this instrumentation to the extent that accelerated development of baggage and passenger screening methodologies now seem worth revisiting. These advances include (1) the invention of a number of ambient ionization methods that are rapid and do not require any sample preparation and (2) the continued development of small, highly capable mass spectrometers to which these ambient ionization methods can be fitted. These advances are detailed in Appendix C.

The recent advances in mass spectrometry ionization methods have yielded processes (such as ambient ionization) that provide mass spectra from materials on solid surfaces in air without sample preparation and in almost real time. These capabilities, some commercially available, could be implemented in trace explosives screening of the external or internal surfaces of baggage using existing commercial mass spectrometers. These new ambient ionization methods have also been implemented on handheld mass spectrometers in research laboratories, a combination that provides high chemical specificity and sensitivity in a small device. Ambient ionization methods can also be used to examine wipes after they have been used to transfer material from baggage in the course of secondary screening. These new methods may offer the sensitivity, speed, and chemical specificity to warrant scrutiny and testing by the Transportation Security Laboratory (TSL) as a possible supplement to or replacement for traditional ion mobility measurements as a secondary baggage-screening device. While additional training may be required, this technology can likely be implemented with limited—if any—additional manpower if it is used to replace the secondary screening that is already in place. If used as a supplement, some additional manpower may be required and this is a trade-off that may need to be considered with regards to the over-all false-alarm rate reduction.

X-RAY DIFFRACTION TECHNOLOGY

X-ray diffraction technology uses energies in the 30 to 80 kiloelectronvolt range to interact with matter, using diffraction and photoelectric absorption to measure the spacing of crystalline materials within the atomic lattice or the arrangement of atoms in a chemical compound. Because the interaction of

BOX 3-1
Potential Value of Mass Spectrometry in Aviation Security Screening According to Previous National Research Council Reports

A 2004 report from the National Research Council, *Opportunities to Improve Airport Passenger Screening with Mass Spectrometry,*[1] has addressed the potential value of mass spectrometry (MS) in aviation security screening. The following recommendations provide an overview of the advantages of this technology:

> TSA should establish mass spectrometry as a core technology for identifying an expanded list of explosives, as well as chemical and biological agents. Specifically, TSA should
>
> - Create a prioritized list of threat materials that are likely to fit a residue scenario and a second list of materials that are not likely to fit the scenario.
> - Determine appropriate MS [mass spectrometry] sampling procedures, inlet configurations, ionization methods, and analysis strategies for relevant materials on this list.[2]
>
> If TSA wishes to improve its trace detection capabilities, it should deploy MS-based detectors in a phased fashion, with successive generations of instruments addressing lower quantities of an expanded list of threat materials and more sophisticated security tasks. These tasks range from passenger screening at checkpoints to monitoring of the air handling system.[3]

[1] National Research Council, *Opportunities to Improve Airport Passenger Security Screening with Mass Spectrometry,* The National Academies Press, Washington, D.C., 2004.
[2] Ibid., p. 6.
[3] Ibid., p. 7.

the energy with the material is chemically specific, some materials contribute more to the false alarm rate than others. X-ray diffraction technology is commercially available and worth consideration as a source of data to help resolve false alarms.

INCORPORATING DATA FROM OTHER SOURCES

In addition to data from explosive trace detection technology, such as mass spectrometry, data from other sources such as carry-on-baggage and passenger-screening checkpoints, perimeter-surveillance data, and even information about passengers' behavior or travel habits can be used to inform the screening process (e.g., selected passengers' bags might be subjected to a more sensitive level of screening.

The 2007 report of the National Research Council's Committee on Assessment of Security Technologies for Transportation, *Fusion of Security System Data to Improve Airport Security*,[2] provides guidance on how best to make use of data from multiple systems (see Box 3-2).

[2] National Research Council, Fusion of Security System Data to Improve Airport Security, The National Academies Press, Washington, D.C., 2007.

BOX 3-2
Systems Approach to Data Fusion

The following material is reprinted from the 2007 National Research Council report entitled *Fusion of Security System Data to Improve Airport Security:*[1]

For the Transportation Security Administration (TSA) to move from the recognition of data fusion as a key technology for transportation security to having an effective plan for implementing data fusion solutions requires a systems approach. This approach would provide the programmatic basis for integrating security systems for checkpoints, checked-baggage screening, and access control. Key outputs from this systems approach that will enable the successful implementation of data fusion are the following:

1. A set of data standards (e.g., Extensible Markup Language [XML]) for the integration of data from security systems and security personnel;
2. A path for the growth and migration of passenger pre-screening as an input to data fusion;
3. Reference frames for exchanging locational data at all levels from within bags to within airports;
4. Standards for the identification of explosives, hazardous materials, and items that appear as hazardous but are not;
5. Common measures of uncertainty for all data inputs and validated error rates from security systems;
6. Data structures for radio-frequency (RF) tagging and other object identification and marking;
7. Ontologies for potential threat objects, systems, subsystems, and scenarios in baggage screening, checkpoints, and airports that enable the linking of alerts, observations, and historical data and provide for dynamic threat assessment;
8. Data structures for airport and airport perimeter kinematics with a particular focus on trajectories;
9. Visualization methods that enable distributed situational awareness and assessment;
10. Standardized data structures for access control, including biometrics; and
11. Standardized data interfaces for access control with facility security.

[1] National Research Council, *Fusion of Security System Data to Improve Airport Security,* The National Academies Press, Washington, D.C., 2007, p. 44.

4

Incentivizing Research and Development to Decrease False Alarms in an Airport Setting

Improvements in technology for reducing false alarms in checked baggage screening in U.S. airports are discussed in previous chapters of this report. The committee believes that, in addition to such improvements, making adjustments in the structure that the Transportation Security Administration (TSA) uses for contracts with equipment manufacturers can lead to advances in technology development and to strengthening the mechanism by which improvements are implemented. Although the discussion in this chapter focuses on for-profit companies that seek to make sound financial decisions about investing their research and development (R&D) funds, academic groups also require long-term planning before establishing research work in a specific area. A long-term strategy for improving explosive detection system (EDS) performance in an airport setting would benefit all of the stakeholders involved and might encourage participation by others as yet not engaged in improving checked-baggage screening.

ADDRESSING CONCERNS OF EXPLOSIVE DETECTION SYSTEM VENDORS

The Need for a Long-Term Transportation Security Administration Plan

The process of developing, testing, and implementing improvements in technology is a long-term investment for a company. Technologists with "a good idea" must convince their management not only that their idea has merit in terms of fixing a known problem or improving the performance of an existing technology, but also that the company will eventually earn back the money that it will spend to bring the idea to fruition and will make some profit on top of that. This kind of long-term planning cannot take place in an atmosphere in which the goals of the potential buyer are unclear or apt to change quickly.

The committee heard from EDS vendors (see the section entitled "Study Process" in Chapter 1 of this report) that the TSA provides them with few incentives to improve the performance of their equipment. Additionally, although the Department of Homeland Security (DHS) aims to improve the false alarm performance of EDSs for baggage screening, the committee was made aware of no clear plan from the TSA to implement improvements in the performance of fielded systems. Vendors variously heard that they should be working on improvements that ranged from reducing false alarms, to reducing operational costs, and even to increasing time between planned or unplanned maintenance events. The result of these mixed signals is that companies may invest in projects that save the companies money but perhaps do not improve the performance of fielded equipment. Without changes to current TSA policy, there will be no incentives for vendors to spend money to develop improvements beyond the necessary fixes for known problems.

Creating incentives for vendors and the technical community to develop improvements will require an organizational framework that includes a known path for the deployment of technology, a realistic strategy for fielding proven improvements, and specific incentives for vendors to provide equipment that performs better than would be necessary to meet baseline requirements.[1] The committee believes that the DHS and the TSA, in cooperation with the equipment vendors, must develop a realistic, long-term strategy for the performance improvement of EDS equipment in an airport setting.

One of the most successful demonstrations of how improvements can be driven by a long-term plan is the semiconductor industry's International Technology Roadmap for Semiconductors (ITRS).[2] This 15-year roadmap was developed through the participation of chip makers, equipment suppliers, and research entities, and over the years it has laid out the generational technology requirements for the industry to continue to realize Moore's law. The roadmap is updated annually by the appropriate teams, which meet each year in a public meeting. In the competitive semiconductor market, this roadmap has served to spur development and manufacturing activities by individual companies and allowed them to remain competitive.

For the DHS and the TSA, a similar approach could result in a consensus on future requirements. Although incentives for participation would be different from those for the competitive market of private industry, and although priorities in a long-term plan involving EDS equipment would necessarily change on the basis of changing threat environments and other outside influences, a long-term plan developed cooperatively would allow companies to evaluate their risk-and-reward strategy in a more stable investment environment. In support of this, the committee re-endorses the following recommendation.

Recommendation: "Within one year, in cooperation with the other stakeholders, the FAA [Federal Aviation Administration] should develop a five-year joint-development plan that includes cost, stakeholder responsibilities, quality measures, and other important factors. This plan should be a living document that is formally updated annually. Buy-in from all stakeholders will be necessary for the plan to be effective." (National Research Council, *Assessment of Technologies Deployed to Improve Aviation Security,* National Academy Press, Washington, D.C., 1999, p. 5.)

Changes Needed for Dealing with Technological Improvements

Technical Review of Changes

A second area in which the committee was made aware of vendor frustration by company representatives was with respect to their realistic expectation that their companies' improvements would be purchased by the TSA for use in fielded equipment. Each vendor that the committee heard from[3] described improvements that could be fielded now but that were being hindered by TSA testing requirements or by a lack of guidance on how to evaluate or implement these changes. As with the long-term strategy, companies will invest in technology improvements that can reasonably be expected to generate a return on the investment. If the company pays for development but then has to wait for the next procurement cycle to see any payback, there is little incentive to improve its product continuously or to evaluate third-party improvements.

The committee does not believe that the TSA should spend money fielding every suggested change. Instead, it should create a framework by which reasonable changes can be evaluated against the claimed improvements and implemented in a sensible way. A first step in that process could be the development of a group of individuals knowledgeable about the technology and with broad experience in the technology, testing, and field requirements (see the section below entitled "Technical Review

[1] Beyond the obvious contracting mechanisms, "incentives" could come in the form of such things as extended patent protection. See, for example, Francesca Cornelli and Mark Schankerman, Patent renewals and R&D incentives, RAND Journal of Economics 30(2):197-213, 1999, which describes an "incentive effect" for R&D that comes from giving firms with R&D capabilities the option of choosing longer patent lives.

[2] Available at http://www.itrs.net/, accessed December 28, 2010.

[3] Representatives of General Electric (GE) Security and of L-3 Communications, presentations to the committee, February 12, 2009, San Francisco, Calif.

Board"). Such technical review boards—with a charter to evaluate potential changes and to identify what testing would be required to ascertain whether a change had the intended effect and what processes would be required to implement the change—are common in industry. A body within the TSA with a similar charter could add some certainty to proposed changes by articulating testing and implementation requirements before money was spent on an idea.

The Need for Testing Capabilities

Following the requirement to determine a path to implementation for a claimed improvement is the need for testing capabilities. For software improvements, such testing might require image data from several hundred bags to demonstrate improved detection; hardware changes might require actually scanning bags in an airport setting to confirm a lower false alarm rate. Each potential improvement would have to be evaluated for risk and reward, and each would require particular testing facilities.

The TSA has a variety of testing abilities now, including the Transportation Security Laboratory (TSL) for certification testing and the TSA's individual laboratories. The TSA can also benefit from the vendors' in-house testing facilities and the availability of realistic explosive stimulants. The recently opened TSA Systems Integration Facility at Ronald Reagan Washington National Airport outside Washington, D.C., may add the capability of doing testing that involves actual passenger bags (as compared to "test sets"). All of these resources should be considered when determining how to test proposed improvements.

The Need to Identify Bottlenecks in the Certification Process

A reduction in the time that it takes vendors to complete the certification process would allow improvements to be more rapidly deployed in an airport setting. To address this, the Transportation Security Laboratory will need to examine its certification process for EDSs with the goal of identifying potential bottlenecks. One approach to this issue might be the development of a method to test systems in an airport setting that operates in parallel with extant systems, allowing data on the same passenger bags within a single airport setting to be compared.

Incentives for Vendors

A third area of change in the TSA's contracting processes would be to provide vendors with the opportunity to receive performance bonuses if their equipment exceeded the required baseline performance. This type of incentive could encourage vendors to work collaboratively with researchers in determining improvements that directly impact the desired performance. The incentive would also make it more attractive for the vendors to seek out third parties that might have research that could lead to a better automated-threat reduction algorithm or other improvement.

To implement such a change, the TSA would have to modify its current contracting approach and determine a method to reward performance that exceeds the baseline and to encourage collaboration. One model might be found in the Department of Defense (DOD), which is moving toward a "performance-based logistics" (PBL) contracting program that creates incentives for vendors to determine the best improvement path and to implement it. The section below entitled "Performance-Base Logistics" describes the DOD approach in more detail.

Finally, the committee believes that the current plan of the TSA to replace all the fielded end-of-life EDSs in a single purchase defeats the goal of continuous improvement and could lock the TSA into years of trying to improve fielded equipment through incremental changes. The ability to purchase new equipment periodically could be a strategic path toward improved performance of EDSs in an airport setting.

This aspect of continuous learning reinforces the recommendation made in Chapter 2:

Recommendation: The TSA should not fund an overall replacement of fielded explosive detection systems, because replacing all the units in service with currently available technology would not allow for learning in an airport setting to inform future performance improvements. Instead, the TSA should plan its capital spending for explosives detection improvements over a period of time sufficient to allow several generations of technology to be to fielded on a limited basis, evaluated, and iteratively improved—thus leading to a gradual improvement in the overall field performance of CT-based explosives detection systems.

COLLABORATIVE CONTRACTING METHODOLOGY

According to information available to the committee, the current contracting methodology utilized by the TSA for airport security equipment employs three types of government funding: procurement, operations and maintenance (O&M), and research, development, testing, and evaluation (RDT&E). In this system, the TSA purchases EDSs from the vendor (procurement) and installs and operates the equipment in the airport (O&M). The TSA also pays the vendor or other contractor to provide equipment maintenance in an airport setting (O&M). If development money can be obtained (RDT&E), then system improvements can be implemented.

Another way of looking at these streams of funding is as follows:

- *Procurement funding* covers the purchase of security systems in limited or full-rate production (e.g., 10 CT systems meeting a given performance specification);
- *O&M funding* covers the original equipment manufacturer (OEM) field service support and the TSA operating costs; and
- *RDT&E funding* covers the new development costs associated with technology investigations, new design activities, and the funding of third-party technology research (done, for example, at universities and laboratories).

The limitations of RDT&E funding in the typical government procurement cycle often severely limit the ability of the TSA to fund new product improvements, because as ideas for new technology insertions emerge from the OEMs and from academia, this form of funding can inhibit continuous process and product improvements. When funding streams are separated in the way that they are in the EDS procurement model, there is little incentive for a vendor to provide equipment upgrades that might improve field performance. From the operational point of view, the TSA does not have money to test and field equipment upgrades that have the potential to reduce false alarm rates or to increase the amount of time between required maintenance events and reduce the failure rate of EDSs—and ultimately reduce operating costs.

This gap in funding for continuous improvement has resulted in frustration on both sides—the TSA cannot always field the best and most-up-to-date equipment, and the equipment vendors cannot benefit from their investments in EDS performance improvements. Changing the approach to procurement and operations could provide the TSA with the flexibility necessary to reap the benefits of investments in performance improvement while offering the vendors an incentive for continuously improving their products. This approach, based on the recent shifts in the DOD procurement process known as performance-based logistics, is described in the next section.

The major incentive that the TSA can offer the equipment vendors to improve the performance of their equipment is through the purchase (procurement) of new products that include improved (more rigorous) systems specifications. Such a process would, by its nature, also require the TSA to have a clear set of defined and measurable standards for performance.

PERFORMANCE-BASED LOGISTICS

Performance-based logistics refers to the purchase of support as an integrated, affordable, performance package designed to optimize system readiness and meet performance goals for a system through long-term support arrangements with clear lines of authority and responsibility. The essence of PBL is buying performance outcomes, not individual parts and repair actions; the contract line item (CLIN) structure is therefore designed around the desired performance.

Under a PBL-based contract, the purchaser (the government) and the provider (the equipment vendor) work together to determine key performance indicators for the equipment, and the purchaser provides incentives for the vendors and other contractors to invest in improvements with a reasonable expectation that these improvements will be evaluated and implemented if successful. This method has been successfully employed by DOD contractors.

Overview of Performance-Based Logistics

The Office of the Secretary of Defense has defined performance-based logistics as "a strategy for weapon system product support that employs the purchase of support as an integrated performance package designed to optimize system readiness. It meets performance goals for a weapon system through a support structure based on performance agreements with clear lines of authority and responsibility."[4]

When employed in the context of the total life cycle of a product, a PBL approach to major system fielding has resulted in superior system performance, operational readiness, and continuous product improvement, which directly impacts incentivized contractor profit. The TSA would benefit greatly by implementing a contracting approach that provides an incentive to the contractor to design and field system improvements that positively impact performance parameters which are determined to be significant indicators of success. In addition to the six steps in the PBL flow shown in Figure 4-1 are the lists of responsibilities of the TSA and the OEM or vendor and the joint responsibilities as suggested by the committee.

Advantage of Performance-Based Logistics

The primary goal of a PBL program is to provide logistics services in a contracting structure that offers incentives for continuous improvement in key measures throughout the life cycle of the product. As implemented by the DOD, the purpose for this contracting structure is to allow the procuring agency and the contractor to select system improvements for implementation that would positively impact the incentivized key measures. The contractor is funded to develop or acquire product improvements, and the government reaps the benefit of higher reliability, improved system performance, improved system readiness, and the implementation of system modifications that accommodate a changing threat level. An example of this is the DOD RQ-7 Shadow Tactical Unmanned Air Vehicle program (Shadow program)

---

[4] ADUSD (Logistics Plans & Programs), Total Life Cycle System Management (TLCSM): Plan of Action and Milestones, updated January 6, 2003, p. 2, available at http://www.acq.osd.mil/log/sci/exec_info/sm_milestone_plan010603.pdf, accessed June 3, 2011.

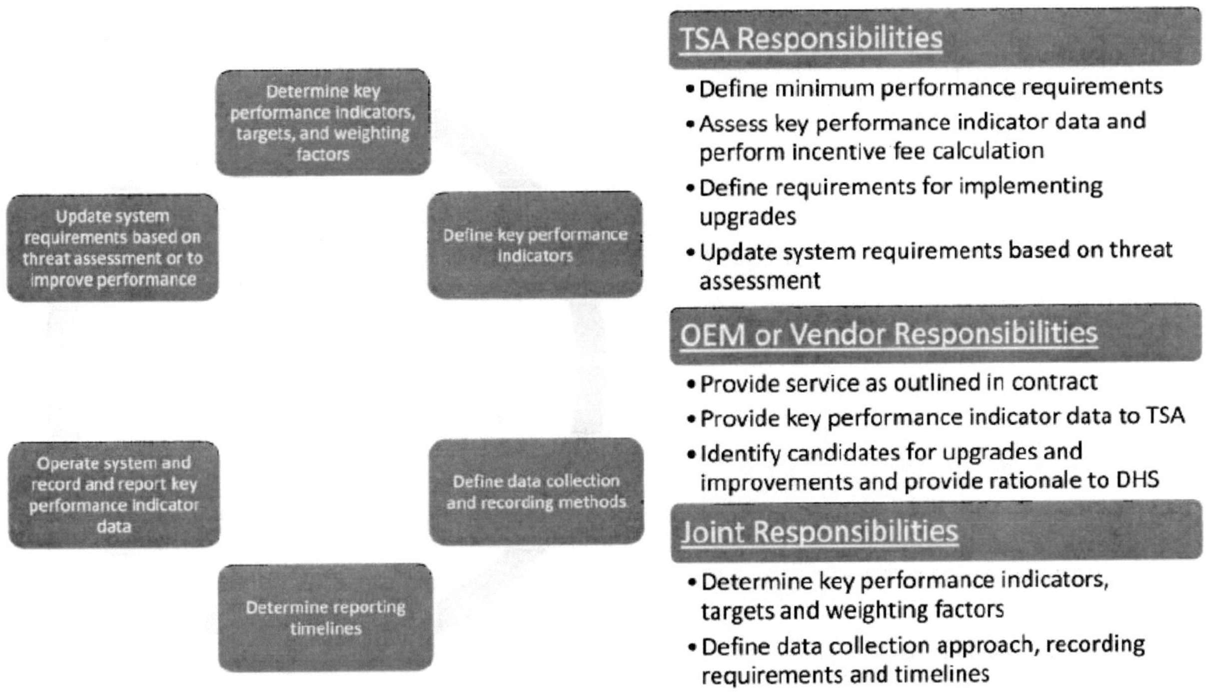

FIGURE 4-1 The steps in a performance-based logistics contract-based flow, and the committee-proposed responsibilities for the Transportation Security Administration (TSA), for the original equipment manufacturer (OEM) or vendor, and for joint cooperation.

that consists of a series of production awards and a companion PBL contract with an incentive plan that has significantly improved system availability and reliability, reduced operating cost per unit, decreased the logistics footprint (inventory and support services), and improved the logistics response time.[5]

Implementation Considerations for Performance-Based Logistics

Many aspects of the PBL process can be applied to the acquisition of and logistics support for airport security screening equipment; one example is outlined in Box 4-1. Like major security systems deployed by the DOD, TSA screening equipment is also vital to the U.S. national defense and addresses evolving threat conditions. Additionally, in both cases, the procurer desires a means to improve its threat-recognition capability continually—be it in an airport, at the airport perimeter, at a train station, at a shipping port, or for the U.S. military on foreign soil.

A typical DOD major system procurement is driven by a statement of objectives that provides the contractor with threshold operational performance requirements. The PBL contract is a "companion" contract (or set of contract line items) that provides life-cycle support for the fielded systems. Performance is measured by a variety of indicators (parameters) that will change throughout the product life cycle, threat, situational environment, and other factors.

As noted earlier, the Shadow program employs a service contract with a fee based on performance metrics that measure results in customer (logistics) support. The customer procures the system, and the contractor provides the full integrated logistics and sustainment support. All spares, repairs, field service representative support, and management are provided under the PBL incentive program. The shadow contract is a cost plus incentive fee contract and is subject to federal acquisition

[5] Performance Based Logistics (PBL) Contract W58RGZ-08-C-0016, U.S. Army Aviation and Missile Command, Redstone Arsenal, Ala.

regulations, DOD directives, and specific contract requirements. Key parameters in the category of system readiness may include minimum (threshold) performance and desired (objective) performance for operational availability, mean time to repair, and mean time between operational mission failures.

TECHNICAL REVIEW BOARD

As noted earlier in this chapter, the committee believes that it would be useful for the TSA to establish a review board of members who represent a broad range of interests for the purposes of evaluating potential improvements and outlining testing and fielding requirements, as well as determining cost of implementation versus potential performance gain. Such a review board would enable vendors not only to establish a stake in the outcome of fielded changes but would also enable them to see a clear path to the implementation of improvements. The board should also review and evaluate methods to identify and mitigate risks, which would assist vendors in making more informed decisions on how to spend their internal R&D (IR&D) funds.

FIELDING CONSIDERATIONS

In addition to establishing a technical review board that could define testing and implementation requirements, the TSA might also establish a capability to review and validate test conditions and results in order to determine whether a specific change meets the criteria set out. Although vendors indicated to the committee that this evaluation is being done, the committee believes that formalizing this role would provide needed structure for decisions that are made with regard to making changes to fielded equipment. This review and validation could be another function of the technical review board, or a separate entity could be established to carry out this function. A testbed for evaluating potential improvements would consist of the following elements:

BOX 4-1
Applying the Performance-Based Logistics Process to Explosives Detection Systems

Below is an example developed by the committee of the types of performance data on explosive detection systems that the Department of Homeland Security might choose to incorporate into a performance-based logistics contract for CT-based explosives detection systems. The numbers are notional only.

Selected Key Performance Indicators

a. System Readiness (Up Time) (25%)
—Assessed periodically and rolled up for all fielded systems

$$\text{System Readiness} = \frac{\text{Up Time}}{(\text{Up Time+Down Time})}$$

b. False Alarms—Current Year (25%)
c. System Maintenance Cost (20%)
d. Reliability Factor (MTBF) (15%)
—The contractor shall achieve a reliability factor goal of 60 days from dock to stock.
—The goal of the metric is to improve the time it takes for depot repair of assemblies.
—This is defined as:

$$\text{Reliability Factor} = \frac{\text{Total days of down-time}}{\text{Number of open and closed work orders}}$$

Operational Reliability Growth Factor (30%)
—Aimed at improving operational reliability by reducing the false alarm rate.
—Contractor and government must plan for investments which will improve the false alarm rate
—Metric defined as:

$$\text{Operational Reliability Growth Factor} = \frac{\text{Cost of False Alarm Resolution (Current Year)}}{\text{Previous Year Investments in System Improvements}}$$ [1]

The minimum and maximum fee table can be determined as:[2]

95-100	15.0%
90-94	13.0%
85-89	10.0%
80-84	7.0%
75-79	5.0%
70-74	4.0%
65-69	3.5%
<65	3.0%

Results will be indexed in a table specific to the parameter, yielding a score for each:
System Readiness (Up Time) = 80
False Alarms = 50
System Maintenance Cost = 75
Reliability Factor = 80
System Manning Cost (including clearing alarms) = 30

Indexing into this fee table yields a fee of 4% of the available fee pool.

Based on the results table and the weights for each performance indicator, a quarterly calculation of performance fee would be calculated as follows:

Incentive Score = (System Readiness Score (80) × 20%)
+ (False Alarm Score (50) × 25%) + (System Maintenance Cost (75) × 15%)
+ (Reliability Factor (80) × 15%) + (System Manning Cost (30) × 25%) = 59.25

[1]A scale must be developed to determine the allocation of points for this metric (e.g., a lower system maintenance cost earns more points, thereby providing incentive to the contractor to institute methods to improve reliability and maintainability).

[2]The assumption is that one year is required to realize the benefit of funds spent on system improvements. The expectation is that the ratio should be greater than or equal to 1.0 in order to justify the investment. A scale must be developed to determine the allocation of points for this metric, depending on the value of the ratio (with a higher ratio earning more points).

1. Access to images from scans of bags within an airport setting,
2. Technical and financial requirement specification,
3. Realistic explosive simulants,
4. Methods to identify and retire risk, and
5. Timely discussions with the evaluation board.

Ultimately this review process should lead to faster certifications at the TSL, which would be to the benefit of both vendors and the TSA.

DATA COLLECTION AND ANALYSIS

In a typical PBL program, the government—with contractor input—establishes the data collection structure, processes, data repository, and training required for the implementation of data collection. For example, in the case of the Shadow program, the government-controlled Unmanned Aircraft System Performance Assessment System is the source-data repository for metric performance evaluation. The contractor participates in the training required to maintain the data collection processes and causes the data collection disciplines to be implemented. The government maintains responsibility for the central data repository and provides the contractor with the levels of data access required to utilize the system for maintenance management, supply chain management, asset visibility, and data analysis. The contractor ensures that accurate data collection and analysis are input into the data collection system to determine metric performance.

PROGRAM MANAGEMENT

The contractor provides management personnel for planning, organizing, scheduling, controlling, and directing all activities in a manner that supports the performance metrics contained in the contract. A PBL program plan is developed by the contractor to address schedules, resources, budgets, and other information required for program management. In addition, the PBL program plan includes management planning, executive management summaries, change logs, functional budget allocation, contract data, program schedules, contract line item numbers, work-breakdown structure, control account managers, organizational charts, procurement planning, subcontract planning, facilities and capital equipment planning, a work-breakdown-structure dictionary, cost performance forecasts, cash-flow schedule, engineering planning, post-deployment software support planning, personnel planning maintenance of action item logs, security and safety requirements, project directives, risk management planning, and various program records.

Box 4-2 describes how the PBL model might be employed in the aviation security setting.

Recommendation: In order to better capitalize on improvements and provide vendors with the necessary incentives to invest in research that will lead to better performance metrics, the TSA should consider adoption of a different contract structure for the procurement and maintenance of the computed tomography-based explosive detection systems used for checked baggage, as well as for other screening technologies. One approach worth considering is performance-based logistics contracting, which is currently used by the Department of Defense.

BOX 4-2
Applying the Performance-Based Logistics Model in an Aviation Security Setting

Suppose that the combination of the Transportation Security Administration (TSA) operating costs and the original equipment manufacturer (OEM) maintenance contract (also known as operation and maintenance [O&M] funding) to cover the O&M of the security screening equipment across domestic airports are approximately $100 million per year.[1] Now suppose, the OEM has a concept for system improvements that would reduce the false alarm rate by 5 percent. This reduction would result in a decreased need for personnel to clear false alarms and a savings of $10 million per year. The OEM has estimated the cost of the design, certification tests, and fielding of the modification to be $17 million, indicating a payback of 1.7 years. Based on the guidelines for selecting technologies for insertion, the TSA decides to fund the contractor to implement this improvement using O&M funding, because the ramifications of the change positively impact the sustainment costs.

A more distant future state of contracting for airport security services might evolve into a fee-for-service arrangement. In this contracting model, the government would own the security equipment and the contractor would operate the equipment. Security concerns might limit the degree to which the government chose to implement this contract arrangement (such as implementing it only in airports with lower threat ratings).

[1] The numbers in this case have been made up to demonstrate how the model would work.

APPROACHES OTHER THAN PERFORMANCE-BASED LOGISTICS FOR PRODUCT DEVELOPMENT AND SYSTEM IMPROVEMENT

Original Equipment Manufacturer Research and Development

Traditionally, original equipment manufacturers have invested corporate profits and/or internal research and development funds for equipment modernization and reliability improvements. This funding is generally very limited, untimely, and difficult to secure. The availability of this type of funding is generally contingent on approval of a business case for recouping the investment through subsequent sales to the government for fielding the modifications. The committee heard from various manufacturers[6] that there have been many instances in which the government had not shown interest in fielding their corporately funded upgrades.

Corporate IR&D funding is generally a component of the burdening structure incorporated into a company's billing rates. The U.S. government recognizes the need to provide contractors with incentives to invest in product improvements, and to the extent that a company can include IR&D in its bid rates (and still remain competitive), this is pre-negotiated with the contractor. Therefore, in situations in which a company has elected to use part of its IR&D funding on new baggage-screening technologies for use in airports, it is to the government's benefit to provide feedback regarding these initiatives so that the companies have some motivation to come to successful conclusion with these investments. A joint long-term development plan between government and industry allows for systematic planning for upgrades to both existing and new technology developments.

[6] Speakers from L-4 Communications and General Electric (GE) Security on February 12, 2009, and speaker from Reveal Imaging on April 28, 2009.

University and Laboratory Research and Development

In parallel with investments by EDS manufacturers, researchers have been studying improvements in automated threat-recognition algorithms at universities, government laboratories, and other industrial companies through private funding, government grants, and other contract sources. This research is not being conducted in coordination with any product development, and the committee saw no structure in place for these researchers to partner with either the government or manufacturers for the testing, evaluation, and, ultimately, fielding of these improvements. With the appropriate incentive system in place, it would be possible to foster continuous improvement by the EDS manufacturers by removing the impediments to cooperation with researchers.

Conclusion: The TSA lacks a structured plan for implementing improved EDSs that would give vendors an opportunity to plan research funding and priorities in accordance with the TSA plan.

Recommendation: The TSA should develop a plan to provide appropriate incentives not only for EDS vendors but also for third parties and researchers in academia in order to improve the overall performance of computed tomography-based EDSs, including their rates of false alarms. Incentives should be provided for both short- and longer-term improvements.

Recommendation: The TSA should develop a long-term strategy for the continuous improvement of performance. Involving all interested parties including EDS vendors and users would increase the probability that all stakeholders work toward the same goals.

5

Lessons from Medical Imaging for Explosive Detection Systems

Imaging technologies are used extensively in medicine for the early detection of disease (screening), for the diagnosis and characterization of disease, for the monitoring of therapy, and for post-therapy surveillance for disease recurrence. These imaging technologies include not only computed tomography (CT) but also more traditional x-ray modalities, magnetic resonance imaging, and various nuclear scanning modalities. The medical uses of imaging have fueled major advances in technology (hardware and software) and in methods of quality control and performance monitoring. An extensive research apparatus, in place for the study of all aspects of medical imaging, has given rise to a large body of scientific literature.

This chapter provides a comparative review of key features of medical imaging systems and explosive detection systems (EDSs). Although the discussion focuses primarily on medical CT because of its technological proximity to CT for EDSs, many of the conclusions apply more broadly to other imaging modalities.

COMPUTED TOMOGRAPHY IN MEDICINE AND IN EXPLOSIVE DETECTION SYSTEMS

Diagnostic radiology was revolutionized with the introduction of computed tomography in the early 1970s[1] because this technology provided two important characteristics: (1) the ability to display relatively high resolution cross sections of human anatomy while (2) assigning quantitative values to the pixels. CT images are scaled in Hounsfield units, which approximately represent the mass density of the object being scanned. No prior diagnostic radiology instrument was capable of quantifying localized tissue characteristics such as density, and thus many studies were initiated to determine the value of this new kind of information. The cross-sectional nature of the images was immediately recognized as a remarkable breakthrough by the medical imaging community, and within only several years, CT was adopted by radiology departments both large and small.

This development fostered intensive research and development by vendors of medical imaging instruments, with relatively few contributions being made by the academic community. Much of the engineering development in medical CT was therefore considered proprietary and has remained unpublished. In the same way, CT-based EDS vendors have sequestered details of their instruments from public knowledge, though with EDSs this is done for security purposes in addition to proprietary concerns.

---

[1] Overviews of the history of medical CT can be found in the following: X. Pan, J. Siewerdsen, P. La Riviere, and W. Kalendar, Anniversary Paper, Development of x-ray computed tomography: The role of *Medical Physics* and *AAPM* from the 1970s to present, Medical Physics 35(8):3728-3739, 2008; E. Krupinski and Y. Jiang, Anniversary Paper, Evaluation of medical imaging systems, Medical Physics 35(2):645-659, 2008; M. Giger, G.P. Chan, and J. Boone, Anniversary Paper, History and status of CAD and quantitative image analysis, Medical Physics 35(12):5799-5820, 2008; and S. Armato III and B. van Ginneken, Anniversary Paper, Image processing and manipulation through the pages of *Medical Physics,* Medical Physics 35(10):4488-4500, 2008.

THE TECHNOLOGY OF MEDICAL COMPUTED TOMOGRAPHY SCANNERS

Modern radiological CT scanners are similar to CT scanners for EDSs. They use rotating cone-beam geometry, with one or two x-ray tubes operating at 80 to 140 kilovolt peak voltage, and the same number of arrays of multi-channel x-ray detectors are arranged on a gantry that rotates at speeds of up to several hundred revolutions per minute. A bowtie filter may be used to reduce the dynamic range and homogenize the beam hardening of the projections. Projection data are typically acquired from many slices simultaneously, so that the entire anatomic volume of interest can be interrogated in under a second. Images are reconstructed by variants on filtered back-projection methods with corrections for scatter, soft-tissue beam hardening, off-focal-spot blurring, detector channel gains, gravity effects on the gantry, tube and/or modulator intensity drifts, and other, more subtle imperfections. The reconstructions typically utilize purpose-built hardware to perform the calculations rapidly. Whole-body scans have reconstructed voxels (volume pixels) of less than 1 cubic millimeter and have selectable algorithms that are optimized for either soft tissue or bone conspicuity.

Each vendor of medical CT scanners initially had a proprietary image file format, but the Digital Imaging and Communications in Medicine (DICOM) format was adopted in 1993 as the industry standard through promulgation by a committee of the National Electrical Manufacturers Association (NEMA). This move was critical, because the standardization allowed third-party vendors of image-processing software and hardware to burgeon, and it provided flexibility to hospitals by equipping radiology departments with a mixture of vendors' devices optimized for their needs.

QUANTIFICATION WITH COMPUTED TOMOGRAPHY

Initially, there was great enthusiasm for exploring the quantitative nature of CT, given that Hounsfield units apparently provide high precision in depicting tissue density. It was hoped that tissue characterization based on these numbers would allow physicians to make informed decisions in detecting and staging pathology as well as for monitoring therapy. For example, studies examined whether a threshold Hounsfield unit boundary value could be set and used to diagnose lung cancer nodules.[2] Unfortunately, this study and others demonstrated that biodiversity precluded this simple level-set approach from being fully successful (the approach was 98 percent sensitive but only 58 percent specific).[3] With this degree of specificity, it was realized that the use of Hounsfield unit numbers alone for cancer screening would lead to unacceptable false-positive rates. Thus, diagnostic CT suffers from the identical problem that plagues CT-based EDSs in that density values overlap between benign and malignant target types.

COMPARISON OF COMPUTED TOMOGRAPHY (CT) FOR MEDICAL USE AND CT FOR EXPLOSIVE DETECTION SYSTEMS

It is already evident that medical CT and CT for EDSs share a number of similarities but also have dissimilarities. It is instructive to examine this comparison in detail to determine if there are opportunities by which medical CT experience can inform EDS design and operation.

First, there is a difference in the nature of the target of detection for the two systems. For diagnostic CT, the type of object being imaged (e.g., a tumor embedded in tissue) is fixed. That is, human pathology does not adapt except through relatively slow evolution or response by mutation to therapies

---

[2] Stephen J. Swensen, Robert W. Viggiano, David E. Midthun, et al., Lung nodule enhancement at CT: Multicenter study, Radiology 214:73-80, 2000.

[3] Sensitivity = PD [probability of detection]; specificity = 1 − PD.

that have been introduced. Thus, the CT characteristics of brain tumors are the same now as when CT was first introduced, although biodiversity guarantees a range of presentations.

By comparison, the "target" for CT scanners used for explosive detection systems is material within a device constructed by humans and purposefully designed to be deceptive in appearance; it can be composed of a wide variety of materials and components and designs that continue to evolve. The nature of these devices therefore can rapidly be altered as their designers adopt new strategies in response to changes in EDS equipment or geopolitical and other stimuli. Because the range of density for many explosives overlaps that of common household materials in checked baggage, the explosives designer has many choices and can readily alter the device composition and disposition as EDSs become more sophisticated. This means that the EDS detection software should be flexible so that changes in algorithms can be rapidly installed in response to the evolution of the explosives threats. That is not the case at the present time because EDSs are supplied and certified as a single package consisting of hardware, reconstruction software, and post-processing software.

Second, in addition to the differences in the nature of the target for the two modalities, there are differences in the conditions of use. Medical CT has traditionally been an imaging modality used in symptomatic cohorts (that is, a test is indicated because of prior clinical findings) and is used to confirm a diagnosis, to determine the stage of a disease, to delineate the site or extent of a pathology, or to monitor the progression of a disease or therapy. The use of CT for imaging asymptomatic cohorts (screening) is more recent and is still in development. For example, CT colonography is under consideration as a modality of colon cancer screening,[4] CT angiography is beginning to be used in the screening and detection of coronary artery disease,[5] helical CT is currently evaluated as a modality to screen for lung cancer,[6] and CT is also under consideration as a modality to screen for breast cancer.[7] The screening uses of CT are still evolving, and as a result the image-processing software is not as much of an intrinsic component of the medical CT scanner as it is of the CT scanner used in EDSs. However, the growing potential of CT as a screening modality is giving rise to the development of imaging software such as computer-aided diagnosis for CT colonography and software for volumetric CT.[8]

A third difference between medical CT and CT used in explosive detection systems is that medical CT relies on human operators to read the scans and render a diagnostic decision. As noted below, an array of software for visualization and classification has been developed for medical CT. However, these systems are not used as a replacement for human judgment. Time pressure to render a diagnosis is not a significant factor except in emergency medical cases. By contrast, EDS use of CT scanners must rely on the automated threat recognition (ATR) algorithm to make the first decision, by nature of the sheer volume of baggage that must be rapidly processed. In order to respond to the compressed time frame in which they must operate, the current EDSs also provide the post-processing algorithms as an inherent proprietary component.

---

[4] C.D. Johnson, M.H. Chen, A. Toledano, et al., Accuracy of CT colonography for detection of large adenomas and cancers, New England Journal of Medicine 359(12):1207-1217, 2008.

[5] G.L. Raff and J.A. Goldstein, Coronary angiography by computed tomography: Coronary imaging evolves, Journal of the American College of Cardiologists 49:1830-1833, 2007.

[6] O. Brawley and B. Kramer, Cancer screening in theory and in practice, Journal of Clinical Oncology 23:293-300, 2005.

[7] K.K. Lindfors, J.M. Boone, T.R. Nelson, K. Yang, A.L. Kwan, and D.F. Miller, Dedicated breast CT: Initial clinical experience, Radiology 246(3):725-733, 2008.

[8] Much of the early computer-assisted diagnostics in medicine can trace its roots to the military's automated target-recognition programs. Such screening and detection programs may also have application to threat recognition in baggage screening. See, for example, John M. Irvine, Targeting breast cancer, IEEE Engineering in Medicine and Biology 21(6):36-40, 2002. Many of the issues discussed by the author (including appropriate ROC settings, the role of the screener, and the problems of missed detections and false alarms) are relevant to an airport setting. Additionally, the military's experience in identifying targets in a cluttered environment and with forces determined to defeat it can inform the aviation security setting.

A fourth difference is that standardization of image output exists only for medical CT. Medical scanners typically stop at simply producing images, without post-processing as an inherent component. As a result, it has been possible to standardize the image file format since the mid-1990s. This has led to a vast array of third-party products for visualizing and classifying images, with input from the academic and business communities key to their development. In addition, scanner vendors have developed workstations with post-processing software for the visualization and analysis of cardiovascular, angiographic, dynamic contrast, and many other functional assessments. Thus commercial pressures have led to a wide range of innovation in the products that are available. By contrast, CT scanners for EDSs at present have built-in, proprietary classification algorithms, and thus only a small fraction of the national expertise in image analysis has been brought to bear on the explosives-detection problem.

In recognition of the success of competition in the medical arena with the adoption of the DICOM file format as the industry standard, a similar plan, the Digital Imaging and Communication in Security (DICOS), has been proposed by the NEMA for standardizing EDS images and allowing competition that includes the academic community for the development of post-processing algorithms. The proposed DICOS standard relies heavily on the work that has already been done in DICOM and adapts those standards for security applications.[9] In the opinion of this committee, this is a welcome and necessary development.

As a fifth difference between the uses of CT for medical and for explosives-detection purposes, false positives and false negatives have different implications for the two modalities. EDS scanners have very strict limits on the time allotted (seconds) for scanning and processing before a decision must be made. Failure to clear a bag in that time means either that the bag is rescanned or that it must undergo manual inspection, both of which can lead to flight delays and passenger inconvenience and have major human resource implications related to the costs in personnel time for resolving false alarms. The requirement of a high probability of detection and the limits in the overall CT examination time augur for caution among those creating the processes and lead to the high false-positive rates by nature of the receiver operating characteristic (ROC) curve.

A sixth difference is that medical CT scanners have dose limitations, but these can be relaxed for CT-based EDSs. This difference enables the use of dual-energy approaches for providing a second degree of freedom in the information available to the ATR algorithm. It is by no means clear that the present design of CT scanners for EDSs is optimized to take advantage of this added freedom.

As a seventh difference, the patient is positioned in the medical CT scanner by a technologist who is trained to do so uniformly and is provided with adequate time for ensuring highly repeatable, diagnostic-quality images. Thus repeated scans are rarely required because of the human precision exercised in the scan setup. With CT-based EDSs, the bags have many styles and shapes, and the bags and their contents are not oriented in a consistent manner when entering the gantry on the conveyor to the scanner. As a result, the same bag scanned repeatedly by a CT-based EDS can have a high degree of inter-scan inconsistency in the images due to image artifacts and finite image resolution; consequently the system might alarm on one scan but not the next. As described in Chapter 3, there is thus an opportunity to improve detection efficiency (to reduce the rates of false positives while maintaining or increasing the true-positive rates) by using repeated scanning of bags.

A final difference is that the development of medical CT scanners has occurred over a period of years, driven by a broad range of consumer needs and marketing preferences. The vendors of these scanners are incentivized by market pressures to deliver systems with the highest image quality and flexible features, and they have developed extensive research and development efforts to remain competitive. This competition is informed by well-publicized academic studies that examine and report in the open literature both the physical characteristics and the clinical performance of the CT machines under clinical (field) operating conditions. The open environment and standardized image format (DICOM) allow easy entrance into this arena to multiple vendors for supplying image-processing and computer-aided diagnosis processing software to the community, where efficacy can be tested and readily

[9] National Electrical Manufacturers Association, NEMA Standards Publication [IC] v01. Rosslyn, Va., 2010.

reported. It also enables open access for research in image classification and analysis by any of the many experienced investigator groups around the world. EDS machines, by comparison, were developed over a much shorter period, in proprietary secrecy, by a small number of vendors, and although not in widespread use until after the attacks of September 11, 2001, these machines had a rapid research and development cycle and, subsequently, mass deployment. The vast general academic imaging community had no part in either the design of the characterization of these instruments, and it remains unengaged.

There are many parallels between medical CT scanners and CT scanners for EDSs, including the equipment development history, the nature of the target or threat to be detected and classified, the time allotted for doing so, the need for automated detection in EDSs, and the difference in operating points on the ROC curve distinguish the two scanner applications. Freedom to increase the x-ray dose in the EDS application and to introduce dual-energy scanning leads to possibilities for additional information recovery for EDSs and may lead to further classification precision. Engagement of the substantial image reconstruction and/or processing community could lead to further beneficial evolution.

Conclusion: The engagement of more members of the academic and industrial communities, as well as of those in the medical diagnostics and military communities having theoretical and applied expertise in image reconstruction and target recognition, could lead to increases in the effectiveness (and, in particular, decreases in false alarms) of CT-based explosives detection.

EXPLOSIVE DETECTION SYSTEMS AND MEDICAL IMAGING

The main points about the comparison of medical CT to CT for EDSs apply more generally to other medical imaging modalities. In particular:

- Screening for disease is the closest analogue to the use of CT-based EDS for baggage screening. However, the time frame for baggage screening is considerably more compressed than that for screening for disease, and the volume of items to be screened is considerably greater. As a result, baggage-screening modalities have higher reliance on automation and software that permit high throughput while maintaining desired accuracy.
- Exposure to radiation (e.g., x-ray, nuclear scans) and other kinds of harm from screening are a significant concern in screening for disease. However, such concerns are far less relevant in the baggage-screening context and permit the use of technology with higher radiation.
- Medical imaging technology undergoes a continuous process of evaluation through formal studies, often comparative, routinely published in the extensive literature on diagnostic imaging. These studies address a broad range of issues and span the developmental trajectory of imaging modalities, from early laboratory and engineering studies to advanced clinical trials evaluating the use of these modalities in a medical setting. The considerable methodological and practical expertise from medical imaging research can be put to good use in fostering the development of a rigorous evaluation of EDS for baggage screening.
- Various approaches have been developed and implemented to monitor the quality and effectiveness of medical imaging in daily medical practice. These approaches are often institution-specific, but they utilize standards and best practices developed and recommended by professional societies and other organizations. For the broadly used modality of mammography for breast cancer screening, a national system of monitoring quality has been in place since the early 1990s. The system was instituted by the Mammography Quality Standards Act (MQSA) of 1992 (P.L. 102-539) and is run by the U.S. Food and Drug Administration. The MQSA regulations include nation-wide quality standards for mammography, with annual inspections, accreditation and certification requirements, standards for reporting results, and requirements for the training, education, and experience of all personnel. Also for mammography, substantial data collection is conducted nation-wide through mammography registries. Data from these registries are used to monitor and evaluate the practice or mammography around the

country. For example, a host of studies conducted by the Breast Cancer Surveillance Consortium have been published in recent years on such topics as the rates of positive mammography findings, the diagnostic accuracy and predictive value of mammography, and factors associated with the variation in positivity rates and diagnostic performance across institutions and individual mammographers.[10]

LESSONS LEARNED

Image Standardization and Post-Processing Software Development

Importantly, the introduction of a standardized image format opened the door to academic participation in post-processing innovations in three- and four-dimensional visualization and computer-aided diagnosis programs,[11] because details of the scanner process were divorced from those details related to the processing of the images for specific applications. This allowed the scanner vendors to retain control over propriety details of the acquisition of images, but it provided easy access to research on the application of these images from a large body of academic and industrial groups with experience in relevant fields of study. Based on the experience with DICOM, there is now a move toward standardizing the image format of CT used for EDSs for the same purpose: to foster participation by academic and other laboratories in the development of post-processing algorithms for explosives detection.

The committee believes that proceeding with the plan to separate the acquisition of CT images from the post-processing programs will improve CT-based EDSs while at the same time inviting greater competition for the development of the post-processing programs. The existing medical image processing field is large and includes dozens of strong academic laboratories as well as well-supported industrial medical research and development programs that have been successful in providing excellent computer-aided diagnosis algorithms. Broader participation by these highly experienced groups with diverse backgrounds in image processing would make it likely that new methods would be developed that may improve the detection and classification efficiency of baggage scanners.

However, one must exercise caution in this endeavor to sever the connection between acquisition and analysis operations. Post-processing success depends on the quality and completeness of the images themselves. It is by no means clear that the image quality is fully appropriate and optimized in current baggage scanners, and thus limiting the availability of the information to only the reconstructed images guarantees that any existing deficiencies in the acquisition and reconstruction processes will not be addressed by the larger community. To enable optimization, a form of raw data (such as sinogram[12]) will need to be made available in a standardized format to a limited community of scientific experts so that they might assess current limitations of CT based EDS images acquisition and potentially derive novel solutions for improved image reconstruction as a part of the problem. Because of the importance of addressing the false-positive rate of CT-based EDS alarms, it will be critical to address the proprietary considerations in a way that allows such a larger community involvement.

---

[10] R. Ballard-Barbash, S.H. Taplin, B.C. Yankaskas, et al., Breast Cancer Surveillance Consortium: A national mammography screening and outcomes database, American Journal of Roentgenology 169(4):1001-1008, 1997.

[11] See, for example, "SecurView Diagnostic Workstations," available at http://www.hologic.com/en/breast-imaging/diagnostic-workstations/, accessed September 12, 2010; and Fang-Fang Yin, Maryellen L. Giger, Kunio Doi, Charles E. Metz, Carl J. Vyborny, and Robert A. Schmidt, Computerized detection of masses in digital mammograms: Analysis of bilateral subtraction images, Medical Physics 18(5, September):955-963, 1991.

[12] In this context, a sonogram is a three-dimensional visual representation of the x-ray signal as it is measured at a specified angle in the imaging plane at varying distances along the detector array.

Quality Control and Performance Monitoring

The experience from the medical uses of imaging provides strong support for the feasibility of and the need for the establishment of nation-wide quality-control standards not only for equipment and processes but also for the training and continuous education of image-interpreting personnel. These standards will need to be based on the results of scientific research in both technology and human factors. The actual performance of the systems in practice will need to be monitored through systematic data collection and analysis, as discussed in Chapter 6 of this report.

Finding: The introduction of an industry-standard medical image format (DICOM) in 1993 fostered the development of a diverse and innovative array of diagnostic and therapeutic image visualization, processing, and automated detection/diagnostic products, fueled by the panoply of academic and private-sector research laboratories with extensive experience in the field.

Recommendation: The Department of Homeland Security should promote the rapid acceptance of a standardized format for EDS images for all TSA-certified machines.

6

Data Collection, Management, and Analysis

BACKGROUND

As described in Chapter 2, current explosives detection and screening practices are conservatively applied at U.S. airports in order that there be reasonable assurance that explosives cannot be placed on airplanes. One result of this conservative practice is the large number of false alarms that result from the screening of checked baggage: According to Transportation Security Administration (TSA) estimates, this larger number of false alarms—also called false positives—translates into hundreds of millions of dollars per year of added cost to the government for resolving the false alarms, as well as causing passenger inconvenience. Although new technologies continue to be developed to improve the TSA's ability to detect and intercept explosives that are intended to damage or destroy commercial airplanes, there is an immediate need to reduce the false alarms associated with checked baggage screening.

One approach to reducing the number of false alarms is to structure a data collection, management, and analysis system to allow studies of screening processes for the explicit purpose of tracking false positives with the intent of obtaining a better understanding of their causes. This clearer understanding of the causes of false positives should facilitate corrective actions for process improvement either with respect to equipment, software, and algorithms or in the area of operator training.

TRANSPORTATION SECURITY ADMINISTRATION DATA

On the basis of the information that it received in meetings and during site visits over the course of this study, the committee understands that the TSA has the ability to collect the following types of data:

- *Baggage-processing data.* These include counts of the number of bags checked, the number immediately cleared by the airport's explosive detection system (EDS), the number of shield alarms (which occur when any area of a bag cannot be penetrated by x-rays), the number cleared by way of the on-screen alarm-resolution protocol (OSARP), and the number cleared at the baggage-inspection room (BIR);
- *The nature of the identified threat that caused a bag to be sent to the BIR.* Examples of such identified threats might be cosmetics; foodstuffs; electronics; books, paper, shoes or other particular materials causing shield alarms; bag parts or a packing style leading to an aggregation error; or non-bag-related causes such as mis-tracking, operator time-out errors, bag jams, or scanner failures;
- *Electronic copies of certain images that cause alarms.* Some airports are currently collecting copies of such images;
- *Results of periodic standard testing on individual EDSs to ensure that they are operating consistently.* Such testing would involve, for example, system voltages and currents and the EDS's ability to detect threats on certain standardized digital inputs;

- *Occasional detailed studies on the baggage-inspection process conducted at a particular location or locations.* Examples of such studies would be those conducted by the National Safe Skies Alliance and Reveal Imaging; and
- *Results from red-team testing.* During such testing simulated threats are inserted into the system to check the ability of the screening system to detect them.

In spite of these vast data collection opportunities, during the course of its site visits the committee was unable to verify that there is any uniform system of data collection and management with the TSA. Without such data, it will be very difficult to manage and improve the baggage screening process.

Recommendation: The Transportation Security Administration should work with the Transportation Security Laboratory to collect and analyze field data in order to characterize the overall performance of the system by computing statistically valid estimates of probability of detection and probability of false alarm for today's CT-based EDSs. These analyses should also be used to better understand the sources of false positives by determining the dependence of these probabilities on material characteristics of potential explosives threats, the variability in the material characteristics, and the characteristics of non-threat materials typically contained in checked bags. These estimates should then be used as baselines for determining the ability of potential improvements to reduce false alarms.

TRANSPORTATION SECURITY ADMINISTRATION DATA MANAGEMENT AND PROCESSING

The collection and organization of baggage screening data will require the development of a special database and data management system that will allow these data to be available for analysis. Procedures for viewing data and extracting relevant parts of the database for special purposes, such as the generation of reports will need to be coupled with this database system. Procedures could also be developed for the extracting of information needed for special studies such as quantitative risk assessments (QRAs), described below, or anomaly detection (i.e., when a sudden change occurs in the "normal" behavior of an EDS in an airport setting). Commercial off-the-shelf software for building such database systems are readily available and reasonably priced, although additional investment in hardware and training will still be necessary.

Finding: Discussion with TSA officials, airport personnel, and vendors indicates some limited-scale data collection and laboratory studies that have enabled the sources of false alarms to be broadly identified. However, system-wide data collection and analysis of the sort necessary to seek out the root causes and guide sustained improvements are not being done.

Conclusion: Without more systematic data on the rates and specific causes of false alarms, the TSA cannot determine what changes are likely to result in reduced false alarm rates and, in fact, do not have the infrastructure in place to determine if an implemented change would result in improved performance.

Recommendation: The TSA should develop and maintain a central database and data management system. The database should contain important historical data, examples of false-positive images, data from previous special studies that have been conducted, and results of periodic standard tests on individual EDSs.

Data from all TSA inspection facilities should be kept in a common format in the form of time-series data that record operating variables in the baggage-screening process from all of the TSA inspection facilities. These records should include frequency counts (in units of bags per hour) of the number of bags handled, the number cleared by the ES, the number having shield alarms, the number cleared by on-screen resolution (OSR), the number sent by OSR to the BIR, and the number of times that the ordnance disposal team (ODT) is called.

The Use of TSA Data in Quantitative Risk Assessment

The interaction between throughput rate, false-positive rate, probability of detecting explosives, human factors, and the probability of an attack suggests the need for continuous, detailed, system-wide modeling and analysis of the TSA baggage-inspection system.

Quantitative risk assessment (see Appendix B for a more detailed description and a simple illustration outlining a QRA for quantifying the cost of false positives) provides methods to study and quantify the risk of extremely rare events, and especially events for which there are very limited data. The thrust of the QRA approach is the quantification of uncertainties, providing a framework for communicating how much confidence one has in reported figures of merit.

However, QRA methods can also be applied to other problems that need to be studied quantitatively, such as the problem of reducing the cost of false positives, or for analyzing the probability of explosives being in airline baggage and being cleared to an airplane. Thus QRA could be useful for assessing potential trade-offs that keep probability acceptably low.

Recommendation: The TSA should employ risk assessment methods to obtain a better understanding of the causes of false positives at both the system and the component level.
QRA could also be an effective approach to analyzing the probability of explosives in airline baggage and for assessing the effects that changes to the baggage-inspection system will have on both probability of false alarm and probability of detection.

These QRA studies could be performed on an as-needed basis to develop the understanding of and to improve the baggage-screening processes. Data analyses should explore trends and possible changes in the false alarm rate over time and assess the real effects of changes to the system and procedures. For example, as noted in Chapter 2, changes in rules and charges affecting airline travel can have an influence on the mix of items in checked baggage, and such changes could have an important effect on inspection operations.

The Use of TSA Data for Process Monitoring

A fundamental principle of process operation is the need to monitor important process variables over time. This monitoring is important for purposes of detecting changes as quickly as possible, for maintaining control of the overall system, for effecting improvements to the process, and for quantifying the effect of process improvements.[1]

Such data analyses should be able to identify unexpected changes in the process and also suggest changes that have the potential to improve the process. The data already collected by the TSA need to be linked to the current criteria for clearing baggage and to other control variables that would have direct impact on the performance of the baggage-inspection system.

[1] Other industries use such methods for process monitoring, and the realm of aviation security may be able to learn from them. For example, the chemical industry uses what is known as statistical process control (SPC) to track critical parameters over time. SPC is most useful when there is a baseline process that is expected to behave in a stable manner over a period of time. For aviation security, SPC may be useful for daily machine calibrations and similar verification activities.

The TSA baggage-inspection process is both complicated and expensive to operate. Information in the proposed database, if used properly, would be useful in helping the TSA to identify weaknesses in its systems, improve the systems' processes, assess the real effect of changes in the systems, and keep inspection processes operating properly.

It would be possible to develop software procedures to generate automatically and inexpensively periodic management-level reports that could provide information on the state of the system and flag significant changes or other potentially interesting findings. Reports could be generated for and sent to individual airports. An overall summary, providing system-wide metrics, could also be included. The content of such reports should be highly graphical, showing trends and patterns in important performance metrics (such as screening cost per bag, probability of false alarm [PFA] and probability of detection [PD]).

The Use of TSA Data for Understanding the Root Causes of False Positives

A detailed, quantitative understanding of the root causes of false positives is important if the TSA is to reduce the costs associated with these false positives without increasing other risks. For instance, the overall false alarm rate includes two distinct "populations" of bags, each of which would require a different approach to reducing false alarm rates:

- The first population includes bags for which the EDS cannot make a decision—so-called "exceptions," such as bags containing solid objects that cannot be penetrated by the EDS x-rays, mis-tracked bags, and bags that are poorly positioned in the EDS in such a way that the EDS cannot interrogate the entire bag ("cut bags"). These exceptions are sent directly to the baggage inspection room without the opportunity for a screener to evaluate the image and clear the bag.
- The second population includes bags whose contents include items that are misidentified by the EDS as potential threat items—for example, when the item's properties fall within the window defined for threat items, or multiple items are mistakenly aggregated into a single object that meets the criteria for a potential threat item.

Without systematic data that can be used to establish how much each population of bags contributes to the overall false alarm rate, or what the specific causes of false alarms are within each population, it is difficult to know what the right course of action is.

Recommendation: The Transportation Security Administration should track broad categories of bags with the goal of understanding how each category contributes to the cost of resolving false alarms.

Categories should include the following: the number of bags scanned, the number of bags declared exceptions, the number of bags declared potential threats by the EDS and cleared by the screener using the on-screen alarm-resolution protocol, and the number of bags declared potential threats by the EDS and sent by the screener to the baggage-inspection room for further inspection. Tracking these data over multiple airports and multiple seasons would give the TSA a better overall understanding of the cost drivers contributing to the false alarm rate.

Although some studies—such as Reveal Imaging's Image Quality Evaluation program—have been conducted and are an excellent start, they have been limited in scope and do not allow for seasonal and regional variation. Ultimately, data collection endeavors must be system-wide. The wide range of false-positive images that current screening practices detect could be partitioned into a manageable number of categories of baggage items (e.g., cosmetics, food-stuffs, or metal) and non-bag related causes (e.g., algorithm issues, losing track of bags during the screening process, or hardware faults). Again, better data from the entire screening process is needed to assess the merit of this approach.

As a TSA database is established, taking the opportunity to gain as much information as possible is important. It would be prudent at this stage to collect too much data rather than too little.[2] Counts of the occurrence of the different root causes of false positives should be included for any TSA database that is developed, along with the other data—by-hour, by-bag, by-airport, and by-standard operational data. Anecdotal evidence about false alarm causes in some airports has been presented to the committee; however, it would also be useful to quantify how the frequency of the different root causes changes—for example, with the season, year, or destination. Knowledge of the frequency of alarms for each of these categories might suggest a further decomposition into more specific articles (e.g., cosmetic gels or liquids, as opposed to the whole category of cosmetics, which includes gels, liquids, creams, powders, and pastes, among other substances) to provide clearer guidance about where the highest payoff for corrective actions lies. For each of these categories, it would be possible to have a link to a set of example images of false positives that lead to false alarms.

The goal of examining data such as those described above would be to identify a category or categories of past false-positive images that, on closer examination, provide a basis for more sharply defined criteria that result in fewer false positives. The criteria for establishing the image categories should be driven by the level of likeness to an explosive image. It may be necessary to perform tests and studies in order to provide a technical basis for explosive image-standards for images in the individual categories: The concept is to identify images for each category that vary from having no likeness to explosives to having varying degrees of likeness. Thus, this task must involve experts in interpreting images of improvised explosive devices to sort out the categories.

Two primary classification methods are used for other forms of indexing: (1) K-means[3] and (2) hierarchical ascendant classification have demonstrated some usefulness in classifying images in the medical and biomedical fields.[4] However, the ultimate choice and means by which this should be accomplished will depend on the nature of the population of images both with and without threats.

Information from such image classification studies could be fed back to automated threat recognition (ATR) algorithm developers and also could be used in focused training for OSR operators. Data-mining communities from other fields such as computer science, medical image analysis, and genomic analysis, among others, might also be able to help inform and guide this process.

Recommendation: The TSA should develop a categorization system to record particular causes of false alarms for baggage sent to the baggage-inspection room. The TSA should develop a database to store this information and use it to monitor performance variation and trends over time.

Other Uses of TSA Data for Process Improvement

Ideally, interactive tools would be coupled with this automatic database system so that researchers could investigate parts of the database not included in the automatically generated reports and could extract potentially interesting slices of data as inputs to other systems (e.g., standard desktop data analysis software). The quantitative risk assessment tools, which are driven by process data and other information, could be used to investigate the "what-if" questions that would be useful for quickly assessing the impact of proposed changes to the baggage-handling system. Then the process-monitoring tools could be used to assess the actual effect on the false alarm rate caused by any changes.

Uniform reporting standards that can be used to generate reports automatically, giving detailed information for each screening facility, would be a necessary part of any data management system that is

---

[2] The size of the sample necessary to be statistically relevant ultimately will be dependent on the level of precision desired, the number of variables considered, and the number of effects being measured.

[3] The data set is split into a given number (K) of subsets so that each subset is maximally compact.

[4] See, for example, J. Frank, Three-Dimensional Electron Microscopy of Macromolecular Assemblies, Academic Press, San Diego, Calif., 1996.

established. All data should be from the permanent database and should be available for analysis and study.

It is expected that this approach to examining the false-positive data and the decision-making processes for clearing baggage, involving both machines and humans, would lead to a technical basis for obtaining more informative on-screen images of items that may or may not involve explosives. The committee believes that such results would provide the TSA and researchers with a technical rationale for changing equipment specifications, algorithms, and detection criteria that should result in the reduction and better management of false positives.

Recommendation: The TSA should develop a system for sharing false-positive data with detection-equipment vendors, including ATR algorithm developers and, when reasonable, with baggage vendors. Vendors should have a clear picture of how well or poorly their own equipment and that of their competitors is operating in an airport setting.

The above approach to data analysis should greatly facility identifying the causes of false positives and the forms of corrective action that might be taken to reduce the number of false alarms. Of course, the analysis has to be repeated periodically to account for changes in technology and the tactics of terrorists. If it turns out that airport variability is important to an understanding of overall system state, different locations may have to be sampled for data processing.

The above approach should provide a basis for corrective actions with respect to those false positives that can be manifested directly from experience data. More sophisticated analytical models may be required to link rare but high-impact false positives to their fundamental origin. That is, such models may be needed to give consideration to the contribution to false positives made by any part of the total checked-baggage-screening system, be it the passenger, baggage design, baggage-handling equipment, individual screening devices, or screener—including the assessment of changes in processes related to human factors such as the use of threat image projection (inserting a pre-set image of a potential threat among the real-time scans to verify the ability of the TSO to recognize threats)—or the process of physical examination of the baggage. An example of a more sophisticated approach is to perform a quantitative assessment of the risk of false positives and the consequences thereof (e.g., they may lead to a false sense of security and cause delays), as well as the potential consequences of a missed detection. An extension of the approach to such an analysis is illustrated in Appendix B.

It is important to use process-monitoring data to gain insight into the how the baggage-inspection process works and how it might be improved. It is also important to conduct special studies in order to assess conditions that will develop in the future in the TSA baggage-inspection system.

The Use of TSA Data from Red-Team Testing

It is essential that there be strong assurance that the probability of detection is being maintained in the complete baggage-inspection process. There is concern that changes in the TSA protocol (specifically, changes made in an effort to reduce the false alarm rate), traveler behavior, local facility conditions, and various uncontrolled factors could have an adverse effect on PD. Red-team testing, based on a standard bag set containing simulated threats that the inspection process would be expected to catch, can be used to study the actual operating characteristics of the complete system in its actual operating environment.

Recommendation: In addition to collecting performance data on a routine basis, the TSA should, from time to time, conduct special studies and experiments for the purpose of obtaining additional information that would be useful for improving the baggage-inspection processes.

It may be important to conduct studies such as those recommended above at multiple locations, as there could be interaction effects between the factors that are being studied and the inspection equipment being used at different locations or the mix of the baggage at different locations.

Recommendation: The TSA should develop procedures for periodic verification to ensure that fielded EDSs meet detection-performance-level standards that correspond to the requirements for EDS certification. In addition to monitoring detection capability directly (e.g., using standard bag sets and red-team testing), these procedures should include the frequent monitoring of critical system parameters (e.g., voltages and currents) and imaging parameters (e.g., image resolution and image noise) to detect system problems as soon as they arise. For purposes of monitoring EDS performance, the TSA and EDS vendors should develop specification limits for all critical system parameters (and their tolerances) that could be monitored frequently and recorded to track changes in performance during normal operations or to verify performance after maintenance or upgrading.

DISCUSSION

The TSA has the potential to collect large amounts of data, and these data contain important information. However, the committee found no evidence that the data are being collected or used effectively. Establishing a database and a data management system would allow the TSA to extract important information from its data, facilitating process control and process improvement. Having a deep quantitative understanding of the root causes of false positives would help with finding ways to reduce the probability of false alarms without lowering the probability of detection.

Methods of quantitative risk assessment, driven by information in the recommended TSA database, would be useful as an assessment and decision-making tool and would help uncover relationships among the many systems inputs and controls and operational costs, as well as help quantify the risk of a harmful attack.

To keep the baggage-inspection process running correctly and to have the tools needed for process improvement, it will be necessary to employ process-monitoring methods that make use of the stream of data being generated by the process and to have detailed knowledge of the root causes of false positives.

Appendixes

A

Biographies of Committee Members

Sandra Hyland, *Chair*, has 25 years experience in program management in both for-and non-profit organizations. She is currently a senior semiconductor engineer at BAE systems. Prior to that, she served in various positions at Tokyo Electron. She has also served as a staff officer at the National Research Council's (NRC's) National Materials Advisory Board and an advisory engineer at IBM. Dr. Hyland has a Ph.D. in materials science and engineering from Cornell University, an M.S. in electrical engineering from Rutgers University, and a B.S. in electrical engineering from Rensselaer Polytechnic Institute. Dr. Hyland is a member of the American Vacuum Society, Electrochemical Society and the Institute of Electrical and Electronic Engineers. She is a fellow of the Society of Women Engineers, and previously served as a vice chair of the NRC Committee on Technologies for Transportation Security.

Cheryl Bitner is vice president for programs at Pioneer UAV, Inc., and is responsible for program execution for Pioneer's unmanned air vehicle programs. Prior to taking her position at Pioneer, Ms. Bitner worked in various capacities at AAAI Corporation, including director of quality systems, and program director for such groups as fire fighter trainers, electronic warfare trainers, maintenance trainers, and on-board (embedded) trainers. She has more than 28 years of industry experience in providing products and services for the Department of Defense and has a strong background in cost- and schedule-control techniques. Her responsibilities include ensuring positive program performance, strategic planning, manpower management, and personnel development. Ms. Bitner is a certified project management professional, certified software quality engineer, and is a member of the American Society for Quality. She has published a cost-and-benefit analysis of piloting and navigational team trainers and contributes to the AAI Training Systems newsletter. Ms. Bitner holds an M.S. in engineering science and a B.S. in computer science from Loyola College and has completed the Advanced Program Management Course at the Defense Systems Management College.

R. Graham Cooks is the Henry B. Hass Distinguished Professor of Chemistry at Purdue University where he has spent the bulk of his career. His interests involve construction of mass spectrometers as well as their use in fundamental studies and applications. Dr. Cooks is a past president of the American Society for Mass Spectrometry and is on the boards of a number of scientific journals; he has been honored by awards from the American Chemical Society and other organizations. His work is highly cited (one of the original 100 most-cited chemists) and he has served as mentor to some 97 Ph.D. students in analytical chemistry. He holds a B.S., M.S., and Ph.D. from the University of Natal in South Africa, and a second Ph.D. from Cambridge University.

Carl R. Crawford is president of Csuptwo, LLC, a consulting company in the fields of medical imaging and Homeland Security. He has been a technical innovator in the fields of medical and industrial imaging for 25 years. His technology has resulted in 79 U.S. patents and approximately $1.5 billion of revenues for his clients. Dr. Crawford was the technical vice president of corporate imaging systems at Analogic Corporation, Peabody, Massachusetts, where he led the application of signal and image processing techniques for medical and security scanners. He developed the reconstruction and explosive detection algorithms for the Examiner 6000, a computerized tomographic (CT) scanner deployed in airports worldwide. He was also employed at General Electrical Medical Systems, where he invented the enabling technology for helical (spiral) scanning for medical CT scanners, and at Elscint, where he developed technology for cardiac CT scanners. He also has developed technology for magnetic resonance imaging (MRI), single photon emission tomography (SPECT), positron emission tomography (PET), ultrasound imaging (U/S), dual energy imaging and automated threat recognition algorithms based on computer aided detection (CAD). Dr. Crawford has a doctorate in electrical engineering from Purdue University and is a Fellow of the Institute of Electrical and Electronics Engineers. He also has adjunct appointments at Northeastern and Virginia Tech Universities.

B. John Garrick (NAE) is an executive consultant on the application of the risk sciences to complex technological systems in the space, defense, chemical, marine, transportation, and nuclear fields. He served for 10 years (1994-2004), 4 years as chair, on the U.S. Nuclear Regulatory Commission's Advisory Committee on Nuclear Waste. His areas of expertise include risk assessment and nuclear science and engineering. Dr. Garrick is a member of Society for Risk Analysis (President 1989-90), and recipient of that society's most Distinguished Achievement Award, in 1994. He has been a member and chair of several NRC committees. He has published more than 250 papers and reports on risk, reliability, engineering, and technology. He has also written several book chapters, and was editor of the text, *The Analysis, Communication, and Perception of Risk*. Dr. Garrick received his Ph.D. in engineering and applied science from the University of California, Los Angeles, in 1968. His fields of study were neutron transport, applied mathematics, and applied physics. He received an M.S. in nuclear engineering from UCLA in 1962, attended the Oak Ridge School of Reactor Technology in 1954-55, and received a B.S. in physics from Brigham Young University in 1952. He is a fellow of three professional societies: the American Nuclear Society, the Society for Risk Analysis, and the Institute for the Advancement of Engineering.

Constantine Gatsonis is a professor of medical science (biostatistics) and founding director of the Center for Statistical Sciences at Brown University. He is a leading authority on the evaluation of diagnostic and screening tests and has extensive involvement in methodologic research in medical technology assessment and in health services and outcomes research. He is group statistician of the American College of Radiology Imaging Network (ACRIN), for which he also serves as a chief statistician both the Digital Mammography Imaging Screening Trial (DMIST) and ACRIN's arm of the National Lung Screening Trial (NLST). A major focus of the research publications and current interests of Dr. Gatsonis is on Bayesian inference and its applications to problems in biostatistics, with emphasis on the evaluation of diagnostic imaging and health services and outcomes research. In addition to Bayesian methods, Dr. Gatsonis has published on other aspects of methodology for the analysis of correlated ROC data and on broader issues of study design in diagnostic test evaluation. Dr. Gatsonis is the founding editor-in-chief of Health Services and Outcomes Research Methodology and an associate editor of the Annals of Applied Statistics, Bayesian Analysis, Statistics and Probability Letters, and Clinical Trials. Previous editorial experience includes membership of the editorial board of Statistics in Medicine, Medical Decision Making and Academic Radiology. Dr. Gatsonis was elected fellow of the American Statistical Association and the Association for Health Services Research.

Gary Glover (NAE) is a professor of radiology and director of the Radiological Sciences Laboratory at Stanford University where he also serves as a professor of electrical engineering. Prior to assuming his positions at Stanford, Dr. Glover was a senior physicist at the GE Medical Systems' Applied Science Laboratory where, in 1985 he won the Steinmetz Award. Dr. Glover is a member of the International Society for Magnetic Resonance in Medicine (ISMRM), the Society of Magnetic Resonance in Medicine, and Society of Magnetic Resonance Imaging, and the American Association of Physicists in Medicine among others. He serves as an editor on the *Journal of Magnetic Resonance Imaging, Medical Physics, Radiology, Journal of Magnetic Resonance,* and the *Journal of Computer Assisted Tomography,* among others. He chairs the NIH Diagnostic Imaging study section, and serves as an ad hoc member on numerous special emphasis sections. Dr. Glover has won several awards for his research and contributions to the field, including: the ISMRM Gold Medal and the Radiological Society of America's Outstanding Researcher Award.

Subhash R. Lele is a professor in the Department of Mathematical and Statistical Sciences at the University of Alberta. He has a Ph.D. in statistics from the Pennsylvania State University. Dr. Lele's expertise is in the statistical analysis of forms and shapes with applications in medicine; and spatial data analysis with applications in public health, ecology, and environmental sciences.

Harry E. Martz, Jr., is the nondestructive testing and evaluation research and development thrust area leader for the Lawrence Livermore National Laboratory. Dr. Martz has extensive background in the use of computed tomography and x-ray radiography to perform nondestructive evaluation. His current projects include the use of noninstrusive x- and gamma-ray computed tomography techniques as three-dimensional imaging tools to understand material properties and to assay radioactive waste forms. Dr. Martz has served on several NRC committees and panels dealing with the general topic of aviation security, including chairing the Committee on Technical Regulation of Explosives Detection Systems.

William Q. Meeker is a professor of statistics and a Distinguished Professor of Liberal Arts and Sciences at Iowa State University. He is a fellow of the American Statistical Association (ASA) and the American Society for Quality (ASQ) and a past Editor of *Technometrics*. He is co-author of the books *Statistical Methods for Reliability Data* with Luis Escobar (1998), and *Statistical Intervals: A Guide for Practitioners* with Gerald Hahn (1991), six book chapters, and of numerous publications in the engineering and statistical literature. He has won the ASQ Youden prize four times and the ASQ Wilcoxon Prize three times. He was recognized by the ASA with their Best Practical Application Award in 2001 and by the ASQ Shewhart medal. He has done research and consulted extensively on problems in reliability data analysis, reliability test planning, accelerated testing, nondestructive evaluation, and statistical computing.

B

Quantifying the Risk of False Alarms with Airport Screening of Checked Baggage

The false alarm rate for checked baggage is high, and these bags must all be inspected by hand, adding a great deal to the overall processing cost for baggage inspection. The Transportation Security Agency and the Department of Homeland Security are seeking recommendations from the National Research Council for actions that will reduce the rate of false alarms, while not unduly compromising the throughput rate of baggage being screened or the probability of detecting explosives. Probability of detection, false alarm rate, and throughput are interconnected, and any solution proposed will result in trade-offs. For example, some actions to decrease the rate of false alarms will lessen the probability of detecting explosives. Thus, there are important constraints on reducing false alarms that must be taken into consideration when making any recommendations for their reduction.

This appendix outlines an approach to quantifying the risk of false alarm scenarios associated with the airport screening of checked baggage and their causes. Studies have been performed on the causes of false alarms and other factors associated with the screening performance.[1] These studies have been based on sampled data from screening operations and revealed the contribution to false positives of different categories of articles such as cosmetics (e.g., creams, gels, powders, lotions).

A rigorous analysis based on the principles of contemporary quantitative risk assessment (QRA) will provide value-added insights for taking corrective actions to reduce the frequency of false alarms. This appendix outlines a systematic process based on QRA principles for a rigorous analysis of the causes of false alarms. The QRA approach outlined here could be extended to allow an informed assessment of the trade-offs in decisions that could reduce baggage-handling costs.

The QRA model proposed tracks baggage through the entire screening process and quantifies the alarm rate at each screening point in a manner that shows the interaction of all components of the screening system. A full-scope QRA would include detailed analyses of the causes of false alarms, which could include hardware, software and algorithms, screening operators, and baggage items and baggage packing procedures. The results of implementing a QRA of the type proposed would be as follows:

- A quantification[2] of the frequency of occurrence of various screening scenarios.
- A quantification of the frequency of different outcomes of the scenarios in terms of the final disposition of the baggage, such as calling in a bomb squad or allowing the baggage to be loaded on the airplane, and the potential consequences of the different outcomes.
- A quantification of the false alarm scenarios burdening the screening system and their causes, thereby enabling the development a roadmap for taking corrective actions to reduce their frequency.

NOTE: This appendix was independently authored by John Garrick, committee member, with the endorsement of the rest of the committee.

[1] Frannie Hamrick, "Field Data for Carry-on and Check Points," presentation to the committee on April 27, 2009, Washington, D.C.

[2] Quantification is taken to mean a full disclosure of what is known about a parameter, including its uncertainty and the supporting evidence. It does not mean absolute certainty, but it implies the quantification of uncertainties.

- A quantification of the uncertainties associated with false alarm rates. The result should be a clearer path forward for a data collection and processing system that more directly exposes false alarms as well as guidance on improved machine algorithms.

The QRA method[3] has been used extensively to enhance the safety and operational performance of complex systems in the nuclear, chemical, aerospace, defense, transportation, and environmental fields. The discussion below is general and omits detailed analysis, because the intent is to describe the QRA approach, not perform an actual QRA.

THE SYSTEM

The quality of a QRA is determined by the extent to which it represents the system being analyzed. In this case the system is a generic screening process for checked baggage typical of U.S. airports. The main components are presented in Figure B-1.

The computed tomography (CT) scanner produces cross-sectional images of the bags. The images are either (1) a set of contiguous slices, known as three-dimensional or volumetric data, or (2) a variable number of slices at varying slice spacing, known as selective slices. The CT scanner may be combined with an x-ray line scanner that is a threat image, projection-ready x-ray scanner, where the images from the scanner are used to determine where to acquire the selective slices.

The automated threat recognition (ATR) system algorithm processes the images produced by the CT scanner to identify the locations of potential threats within bags. Cleared bags (bags with no identified threats) are sent to the airplane. The ATR algorithm is characterized by its probability of detection (PD) and its probability of false alarm (PFA). The ATR algorithm may run on computers in the CT portion of the explosive detection system (EDS) or on the baggage viewing station.

The ATR algorithm also analyzes the images of the bags to determine whether a threat could be shielded from the x-rays used in the EDS. If shielded regions are found in the bag, the bag and its images are sent directly to the baggage-inspection room BIR. If in the course of the baggage handling there is loss of identification of a bag (mistracking), that bag is also sent directly to the BIR.

A computer monitor at the on-screen resolution (OSR) station displays images of bags that the ATR identified as containing potential threats. A transportation security officer (TSO) may clear the decision of ATR using available protocols. Cleared bags are sent to the airplane. If the ATR identifies multiple potential threats, the TSO may clear some or all of the threats.

The BIR receives bags that have not been cleared during OSR. TSOs in the BIR visually inspect the threats or apply explosive trace detection (ETD). If the TSO clears the threats, the bag is sent to the airplane. Bags with remaining threats are handled by an ordnance disposal team (ODT).

Thus, false alarms of threats are driven by both machines and humans. In a machine-driven false alarm, the screening algorithm signals an alarm when there is no threat. Common causes of machine driven false alarms are non-threat substances mistaken for a threat substance or items that aggregate several non-threat items into single items that meet the screening criteria for a threat. A human-driven false alarm involves a TSO. When prompted by the ATR to investigate a specific item or area of a bag, the TSO may mistake a non-threat substance for a threat. In particular, a search for causes of false alarms must investigate the machine, including the data being processed and the ATR algorithm, and the decision-making process of the TSO.

[3] B.J. Garrick and Robert F. Christie, Quantifying and Controlling Catastrophic Risks, Elsevier, Amsterdam, 2008, pp. 17-31.

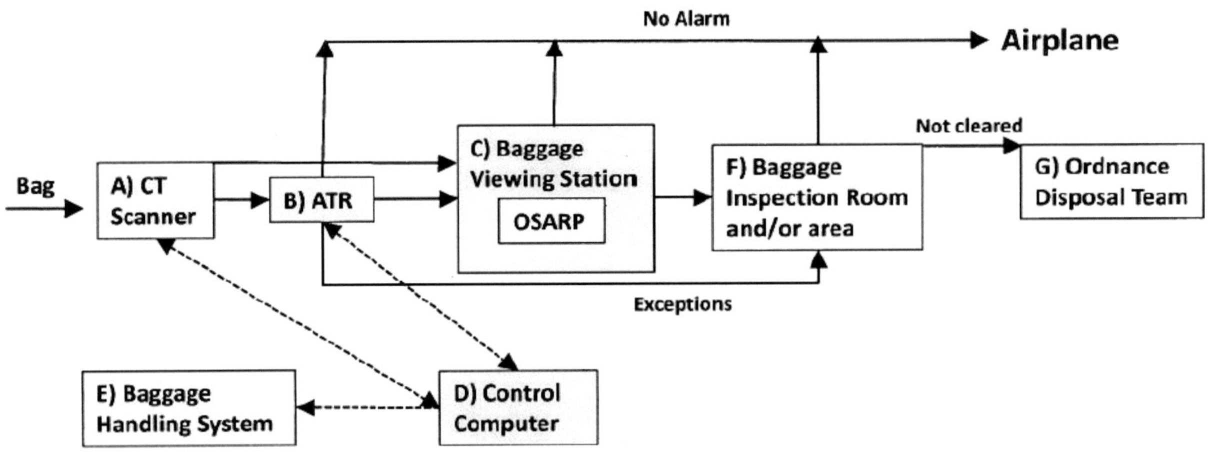

FIGURE B-1 Diagram of an in-line EDS consisting of (A) a CT scanner;(B) an automated threat recognition (ATR) algorithm, (C) a baggage viewing station and the on-screen alarm-resolution protocol (OSARP), and (D) a control computer. This is integrated with (E) the baggage handling system, (F) the baggage inspection room and/or area, and (G) the ordnance disposal team. Shaded boxes are components of EDS. White boxes are subsystems used in conjunction with the EDS. Solid connecting lines show flow of bags and/or images of the bags. Dashed connecting lines show the flow of control and information.

THE MEASURE OF RISK

Generally risk is assessed with respect to a threat to human health (injuries and fatalities), damage to a facility, a transportation accident, an environmental impact, a catastrophic event, or other such situations. In this illustration the committee focuses on the risk of false alarms from EDSs, and in the process exposes their causes to guide corrective actions for their reduction. The parameter of the model is the frequency of false alarms and, more particularly, the frequency with which alarms lead to different action states such as extra screening or even the need for an ODT. Of course, there will be variability and uncertainty in the frequency, and that is something that must be a part of the quantification process. We account for uncertainties using the language of probability.

DEVELOPMENT OF THE RISK SCENARIOS

The cornerstone of the QRA approach advocated here is the triplet definition of risk.[4] That is, when we ask "what is the risk of something?" we are really asking three questions,

- What can go wrong?
- How likely is it to go wrong? And,
- What would be the consequences?

The main task of the risk triplet framework is developing the "what can go wrong?" scenarios. These scenarios can be structured in a variety of ways; the one that should be used is the one that works best for the analyst and the system being analyzed. One framework for structuring scenarios has had a great deal of success: the *event tree*. Basically, an event tree is an inductive reasoning logic diagram that traces the response of a system to different stimuli, that is, to different initiating and intervening events

[4] Ibid., pp. 18 and 19.

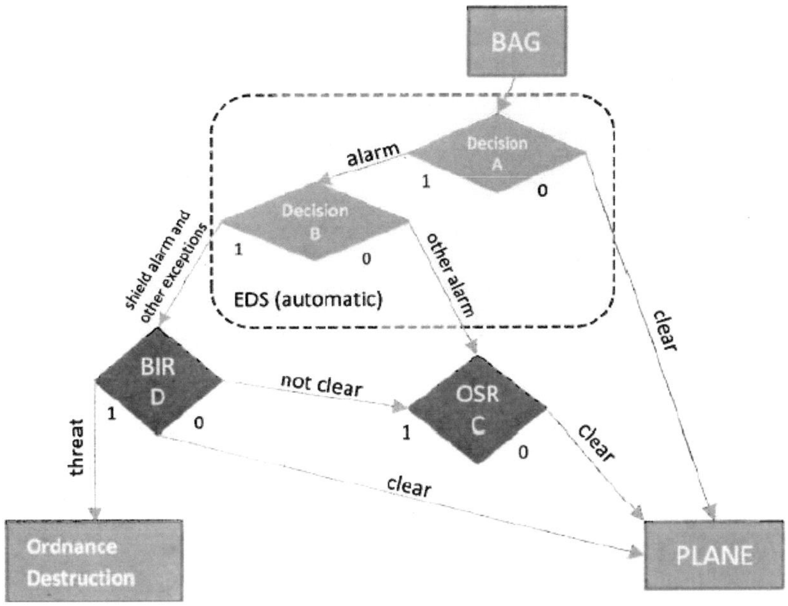

FIGURE B-2 Event-sequence diagram for airport checked baggage.

until the response is terminated in the system either when the system corrects itself through automated or administrative action or when it goes into a particular damage state or undesirable state. Of course, there can be many paths through an event tree. Each path is considered a scenario.

Figure B-2 is an event sequence diagram that outlines the bag-inspection process flow and process logic. Each path through the diagram corresponds to a "what can go wrong?" scenario.

The event tree in Figure B-3 provides the structured set of scenarios desired. It communicates the logic of the system including the branch points and the interaction details of the subsystems A, B, C, and D. Figure B-3 can serve as a roadmap for identifying the various scenarios leading to false alarms. Each path through is a scenario. These paths are readily identified in the figure, where one can see where a scenario originates, which path it takes, and its end point in terms of impeding the flow of baggage to the airplane. For most systems, there are several initiating events and thus several event trees and in some cases thousands or even millions of scenarios. The good news is that the physics of the process usually leads one to a reasonably small set of scenarios that dominate the overall risk.

A comprehensive risk assessment of false alarms would most likely involve segregating the causes and developing separate event trees for each cause set. Candidate cause sets for baggage inspection include cosmetics, foodstuffs, metals/electronics, paper (including books), shoes, bag parts, hardware characteristics, software characteristics, and traveler packing practices. It is obvious that once the logic is laid out and it becomes clear what is needed to quantify the scenarios, the data can be analyzed to quantifying the probabilities associated with each scenario. With knowledge of the scenarios and the attendant logic, the data analysis can be very efficient because it is clear just what information is needed.

QUANTIFICATION OF THE SCENARIOS

The branch points of the event tree are called "split fractions." In order to track the fraction of alarms that are indeed false, it is necessary to know how that fraction varies in order to represent the variability and uncertainty involved. When data exist on false alarms at the branch point and a researcher knows the variability of those data, then the split fraction distribution can be obtained directly from the data. Often such data are limited, and it is necessary to perform some probability analyses to obtain the split fraction distribution that reflects both the variability and the uncertainty. In other words, the available data, other quantifiable information, and probability methods are the basis for assigning distributions to each of the split fractions.

Rates and split fractions vary by such factors as airport, plane destination, time of day, and season of the year. It will be necessary to do separate analyses for given levels of these factors in order to assess—for example, the potential consequences of process or practice changes at a particular class of inspection facilities. For other questions, the scenario distribution can reflect the combination of variability and uncertainty in the occurrence of the scenario. For example, human factors could be captured by probability distributions describing operator responses over a range of image types or bag features. That is, for a given set of bag features, there is a particular probability that the operator clears the bag.

When the data are strong, the scenario distributions would represent only the variability in the frequency of the split fraction. If the data are not strong, then Bayesian methods or other probability methods can be used to process the available data and other information into probability distributions reflecting available knowledge about the split fractions. Because there are only limited data available on the detection of actual threats, these other methods would be needed to assess scenarios in an expanded QRA that assesses the risk of both false positives and false negatives.

FAULT TREE ANALYSIS

To develop the split fraction distributions so as to reveal false alarm causes, the committee introduces another risk assessment tool known as the "fault tree." Whereas the event tree is basically an application of inductive logic and thus the framework for structuring event sequences or scenarios, the fault tree is based on deductive logic and is useful for quantifying split fractions. The fault tree starts with the undesired event—for example, a false alarm—and works backwards decomposing the logic to basic causes or events. Even if it is possible to obtain the split fraction frequencies and their variability directly from field data, fault-tree-type analyses are necessary to reveal the basic events triggering the false alarm. The power of fault tree analysis is the ability to trace undesired events to such basic causes as equipment components and parts, software and algorithms, or human reliability. Figure B-3 is a simple fault tree to illustrate some of the key logic gates used in structuring fault trees. Extensive software exists for processing fault trees.

Complex systems require a more comprehensive set of logic gates than is shown in Figure B-3. Examples of other logic gates include "conditional," "inhibit," and "external event" gates. But Figure B-3 presents the general idea. For example, the diamond-shaped box indicates that the logic of the event remains to be developed. Thus, if equipment were the cause of the alarm the logic would be developed to expose the equipment basic event(s) that caused the false alarm, whether they be hardware failure or a fundamental limitation of the equipment. The task then is to seek the supporting evidence for assigning a probability distribution to that basic event. The fault-tree logic is the equation path for then calculating the probability distribution of the split fraction.

QUANTIFICATION OF EVENT FREQUENCY

Once the split fractions are quantified, the logic of the event tree can be implemented. For example, with respect to the six scenarios in Figures B-3 and B-4 involving the top events A, B, C, and D, and the initiating event, I, the Boolean expressions for scenario, S_i $i = 1{,}6$ are given in Table B-1, where the bar above the letter denotes a threat, i.e., an alarm.

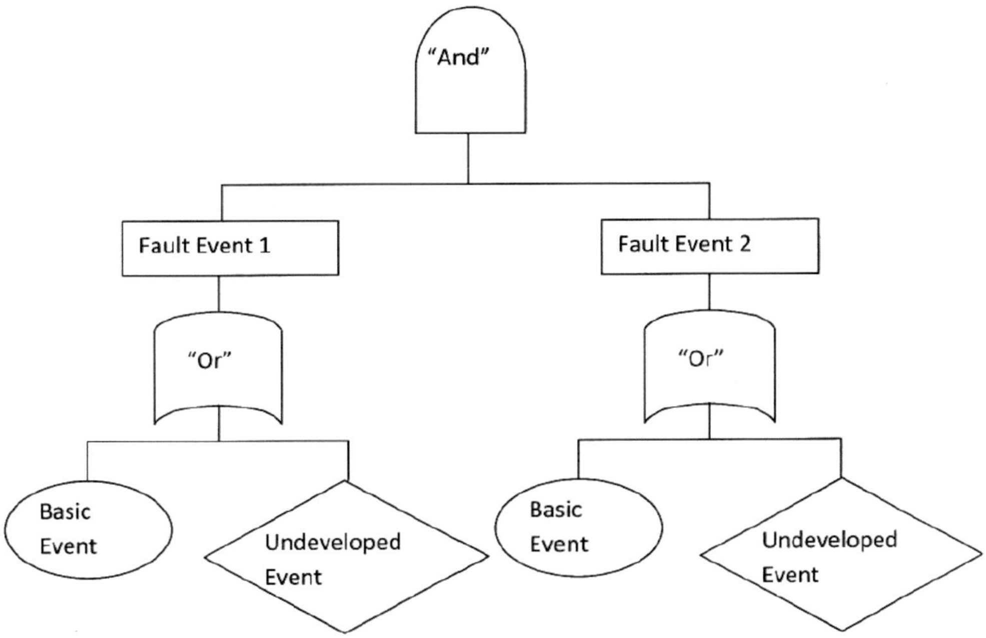

FIGURE B-3 Generic fault tree illustration.

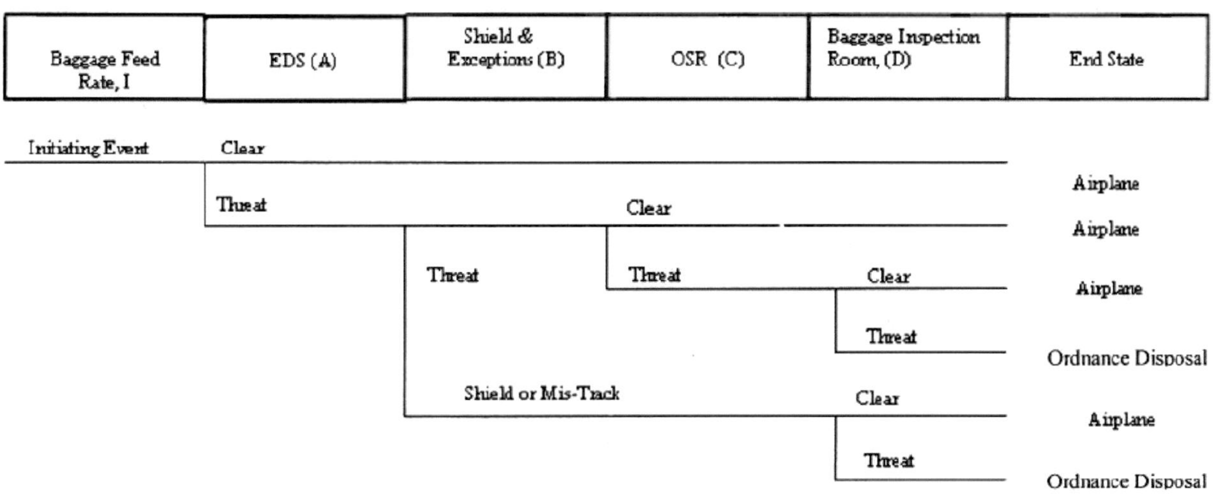

FIGURE B-4 Event-tree diagram for airport screening of checked baggage.

TABLE B-1 Boolean Expressions for Scenarios 1 Through 6

Scenario	Event Description	Frequency	Scenario Cost
$S_1 = I_{ABCD}$	Bag cleared by EDS	$\Phi(S_1)$	C_1
$S_2 = I_{\bar{A}BCD}$	Bag cleared by OSR	$\Phi(S_2)$	C_2
$S_3 = I_{\bar{A}B}\bar{C}_D$	Bag cleared by BIR	$\Phi(S_3)$	
$S_4 = I\ ABCD$	Threat declared by BIR	$\Phi(S_4)$	C_4
$S_5 = I\ \bar{A}\bar{B}C\bar{D}$	Bag cleared by BIR after shield alarm/exception	$\Phi(S_5)$	C_5
$S_6 = I\bar{A}\bar{B}C\bar{D}$	Threat declared by BIR after shield alarm/exception	$\Phi(S_6)$	C_6

Ultimately we want to quantify the frequency with which each scenario will occur. Therefore we need to transform the Boolean equations into frequency equations. Of course, the frequencies of the scenarios and the end states are linked to the throughput rate of the baggage. If the baggage throughput frequency is denoted by I (bags per unit time), the frequencies of the various scenarios, denoted by φ (S_i), $i = 1,6$, are given by the following equations, where the f ($X \mid Y$) s are the split fractions (conditional densities) of X at any given branch point, conditional on the path history in Y:

$$\varphi(S_1) = \varphi(I) f(A \mid I) f(B \mid IA) f(C \mid IAB) f(D \mid IABC)$$
$$\varphi(S_2) = \varphi(I) f(\bar{A} \mid I) f(B \mid I\bar{A}) f(C \mid I\bar{A}B) f(D \mid I\bar{A}BC)$$
$$\varphi(S_3) = \varphi(I) f(\bar{A} \mid I) f(B \mid I\bar{A}) f(\bar{C} \mid I\bar{A}B) f(D \mid I\bar{A}B\bar{C})$$
$$\varphi(S_4) = \varphi(I) f(\bar{A} \mid I) f(B \mid I\bar{A}) f(\bar{C} \mid I\bar{A}B) f(\bar{D} \mid I\bar{A}B\bar{C})$$
$$\varphi(S_5) = \varphi(I) f(\bar{A} \mid I) f(\bar{B} \mid I\bar{A}) f(C \mid I\bar{A}\bar{B}) f(D \mid I\bar{A}\bar{B}C)$$
$$\varphi(S_6) = \varphi(I) f(\bar{A} \mid I) f(\bar{B} \mid I\bar{A}) f(C \mid I\bar{A}\bar{B}) f(\bar{D} \mid I\bar{A}\bar{B}C)$$

As mentioned above, if the data are strong (as might be expected for S_1, S_2, S_3, and S_5, because these events occur frequently), there will be a little uncertainty in the scenario frequency, and the corresponding φ (S_i) might be adequately represented by a single number. Otherwise the uncertainty is quantified with a probability distribution.

Suppose that we wanted to know how frequently false alarms resulted in involvement of the ODT. To obtain this result we would have to quantify the frequencies of both S_4 and S_6. For example, to quantify the frequency of S_6, we need to convolve the distribution of the complete scenario. Figure B-5 illustrates the process whereby the probability arithmetic is usually performed using Monte Carlo sampling methods. The result is a distribution quantifying the uncertainty in φ (S_6).

$$S_6 = I\bar{A}\bar{B}C\bar{D}$$

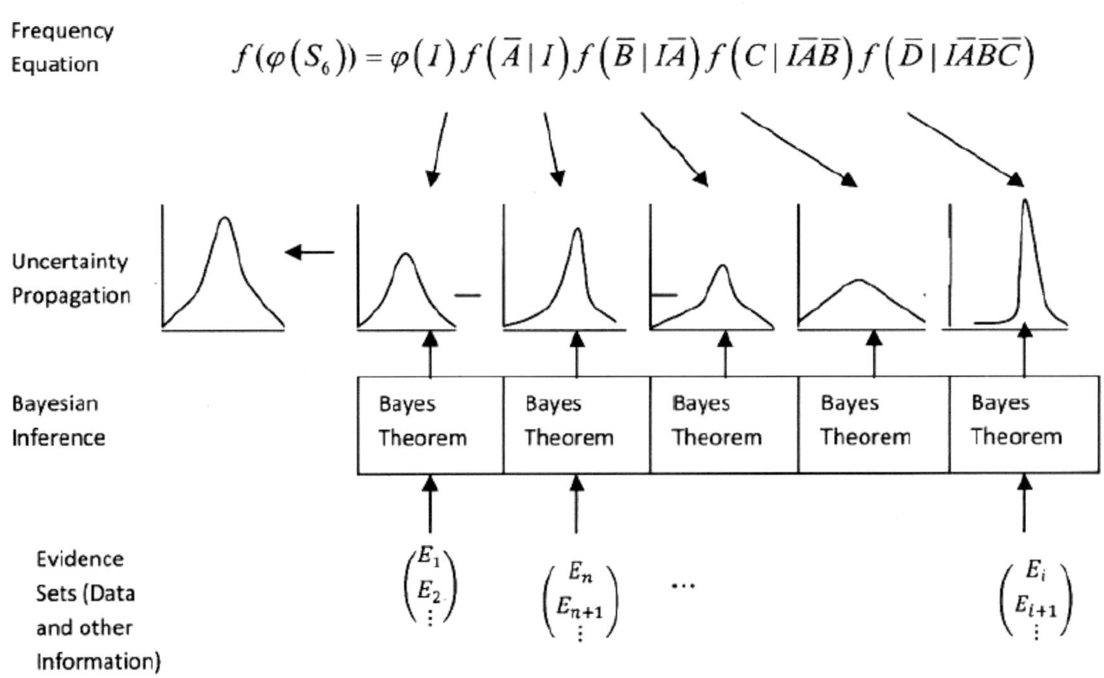

FIGURE B-5 Bayesian convolution of split fraction uncertainties.

ASSEMBLING THE RESULTS

Each scenario has a probability density curve like that illustrated in Figure B-6. The total area under the curve represents a probability of 1. The area under the curve between any two values of φ is the probability that φ (S_6) is between those values.

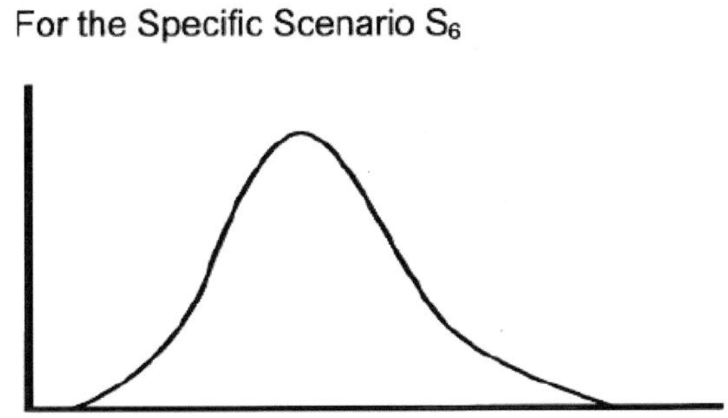

FIGURE B-6 Probability density for frequency φ.

There are several ways to communicate uncertainty in the risk when the uncertainty is quantified by a probability distribution. One approach is to compute a 90 percent probability interval such that the probability (the area between φ_1 and φ_2 of Figure B-7) is 90 percent of the area under the curve. The way to read this result is we are 90 percent confident that the false alarm rate is between φ_1 and φ_2.

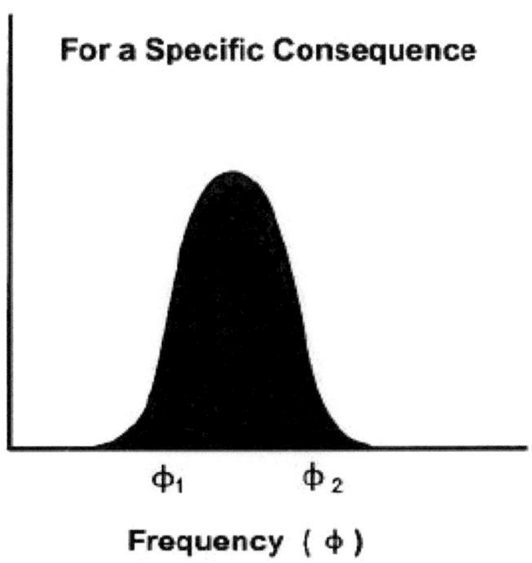

FIGURE B-7 Probability density function.

Probability distributions similar to Figure B-7 can be computed for any given single scenario or for combinations of scenarios leading to a single consequence. For example, the event S_4 or S_6 corresponds to the event that a bag will result in the need to call the ODT. A process similar to that depicted in Figure B-4 would have to be used to convolve φ (S_4) and φ (S_6) to obtain a probability distribution to describe the variability and uncertainty in the frequency with which the ODT needs to be called.

USE OF QRA TO REDUCE RISK

For some decisions the expected cost would be a useful metric for assessing the benefits or the effect of making changes to the baggage-inspection system. For the example in this section, the expected cost could be computed as

$$\text{E(Cost)} = \sum_{i=1}^{6} C_i \text{E}[\phi(S_i)] = \sum_{i=1}^{6} C_i \int \phi(S_i) f[\phi_i] d\phi_i$$

where $f(\phi_i)$ is the probability density function of $\varphi\ (S_i)$.

OTHER USEFUL INFORMATION FROM A QRA

Although risk measures such as those mentioned earlier in this appendix are useful for decision making, they are not necessarily the most important output of the risk assessment. Often the most important outputs is the exposure of the detailed causes of the risks—a critical requirement for effective risk management and process improvement. The contributors to risk are buried in the results assembled to generate the curve in Figure B-6. And in particular, understanding the root causes of risk behind the split fraction analyses illustrated earlier using the fault tree methodology can provide insight to help determine the parts of the system that can be modified to produce important improvements—for example, a modification that will reduce the probability of a false positive without reducing the probability of detecting explosives.

INTERPRETING THE RESULTS

As indicated earlier, the link between the proposed model and the actual reduction of false alarms is the quantification of the total screening process in such a manner that part of the output is the exposure of detailed causes of the risks—in this case, the causes of false alarms. Knowing the causes is the prerequisite for taking corrective actions to reduce their occurrence.

The simplicity of the screening system and the experience base provides an opportunity to collect high-quality data on false alarms. Thus, the most important contribution of applying QRA principles may be to define and fine tune the data management system that will best reveal the causes of false alarms. It is not apparent, however, that any such focused data management system exists that has the scope to quantify false alarm rates and their causes in relation to the total screening system. While the screening process as a system is indeed simple, the individual components of the system are state of the art and are being pushed to their performance limits, primarily because of the throughput demands of airport screening. Experience indicates that applying QRA methods to the analysis of complex equipment has excelled in exposing the fundamental causes of undesired performance outcomes.

EXTENSION OF QRA TO THE ASSESSMENT OF A BOMB THREAT

There are a number of other ways in which QRA might be applied to study and improve the baggage-handling processes. For example, the approach described above was entirely conditional on having no real threats in baggage (the usual situation). The QRA model could be extended to include a parallel set of scenarios conditional on there being real explosive and other components of a bomb in the inspected baggage. The more difficult parts of this extension are the quantification in the absence of any meaningful data of the costs of the scenarios that involve on-board explosives and of the probability of some of the elementary events. This extended QRA would be valuable to assessing combinations of changes to the baggage-inspections process. For example, there might be some changes that would decrease PFA substantially with only a small increase P_d, in combination with other changes that would increase PFA marginally but decrease PD substantially, resulting in a net improvement in PFA without an overall decrease in PD.

C

Chemistry-Based Alternatives to Computed Tomography-Based Explosives Detection

SOME LIMITATIONS OF CT X-RAY METHODS

Methods of detection for explosives based on computed tomography (CT) are fundamentally imaging—rather than chemical—analysis and provide very little molecular information. The measurement provides low-dimensional, cross-sectional images, and the detection depends mainly on density, which is a single number. While some variants of x-ray methods such as dual-energy systems provide elemental information, that information remains limited and the technology has not been widely adopted. It is conceivable that one could manipulate non-explosive materials to get the same output as given by an explosive.

GENERAL REQUIREMENTS FOR ALTERNATIVE TECHNOLOGIES TO COMPUTED TOMOGRAPHY-BASED EXPLOSIVES DETECTION

Any technology developed to augment or replace CT for explosives detection must be molecule-specific because explosives have a wide variety of chemical structures, some of which are very similar to an even larger class of non-explosive materials, and may, moreover, be present in mixtures that have responses different from those of pure explosives. Additional criteria for such an alternative technology are these:

- High sensitivity,
- Rapid analysis,
- Greater specificity than merely sensing,
- Capable of being deployed in a standoff manner
- Employing automatic (rather than manual) sampling

ALTERNATIVES TO CT X-RAYS

Among the technologies that meet some, if not all, of the criteria listed in the preceding section are the following:

- *Dual energy CT.* Offers a second cross section that provides both density and atomic number but not molecular information
- *Neutron activation, x-ray fluorescence.* Provides elemental information, but not molecular information; it can be deployed in a stand-off manner and can penetrate objects.

NOTE: This appendix was independently authored by Graham Cooks, committee member, with the endorsement of the rest of the committee.

- *Ion mobility spectrometry (IMS).*[1] Is attractive for its sensitivity, simplicity, ruggedness, and reliability but is limited in terms of the quality of the molecular information provided and consequently in molecular specificity. More recent work allows explosives recognition by library comparison but still not with high specificity.[2] There is still a great need for approaches that are more molecule-specific, and the combination of IMS with mass spectrometry is one possibility. Supplementary data gained from treatment with reactive gases is another.

Mass spectrometry (MS) and tandem mass spectrometry[3] are another option, which is discussed in greater detail in the section that follows.

MASS SPECTROMETRY

MS is a method of determining the masses of particles, the elemental composition of a sample or molecule, and the chemical structures of molecules. MS does this by ionizing chemical compounds to generate charged molecules or molecular fragments and then measuring their mass-to-charge ratios. In a typical MS procedure,

1. A sample is loaded onto the MS instrument and undergoes vaporization,
2. The components of the sample are ionized, which results in the formation of charged particles,
3. In an analyzer, electromagnetic fields separate the charged particles on the basis of their mass-to-charge ratio.
4. The charged particles are detected, usually by a quantitative method, and
5. The charged particles' signal is processed into mass spectra.

Instruments for performing mass spectrometry consist of three modules:

- An ion source, which converts gas-phase sample molecules into ions.
- A mass analyzer, which employs electromagnetic fields to sort the ions on the basis of their masses.
- A detector, which provides data for calculating the quantity of each ion present based on the measured value of an indicator amount.

The potential of MS in aviation security applications was described in the 2004 report of the NRC Committee on Assessment of Security Technologies for Transportation, *Opportunities to Improve Airport Passenger Screening with Mass Spectrometry.* In that report the committee's focus was MS's potential in explosives trace detection (ETD) to resolve false alarms raised by explosives detection systems. The limitations and advantages of this technology—as well as advances that have taken place since that report's publication—are described in the sections that follow.

---

[1] G.A. Eiceman and Z. Karpas, Ion Mobility Spectrometry, 2nd ed., CRC Press, Boca Raton, Fla., 2005.

[2] See, for example, D.S. Levin, R.A. Miller, E.G. Nazarov, and P. Vouros, Rapid separation and quantitative analysis of peptides using a new nanoelectrospray-differential mobility spectrometer-mass spectrometer system, Analytical Chemistry 78:5443-5452, 2006; and R.W. Purves, R. Guevremont, S. Day, C. Pipich, and M.S. Matyjaszczyk, Mass spectrometric characterization of a high-field asymmetric waveform ion mobility spectrometer, Review of Scientific Instruments, 69:4094-4105, 1998.

[3] In tandem mass spectrometry the ions are subjected to two or more analyses, separated either by space or time.

Some Limitations of Mass Spectrometry

Mass spectrometry for explosives detection is attractive in principle when coupled with new ambient ionization methods, for tandem mass spectrometry (MS/MS) and a small, possibly even handheld, mass spectrometer.

However, the technology also has a number of limitations:

- It is not normally deployed in a stand-off manner,
- The trace analysis method is not suited to bulk examination,
- Representative sampling is very difficult,
- It may be possible to defeat the system by using overwhelming chemicals,
- Quantitation of solids is difficult because of the need for an internal standard, and
- Absolute quantitation methods are not very successful.

One question is where to best apply the technology. Opened bags sent for secondary screening may be most appropriate.

In spite of these limitations and open questions, there have been very significant advances in MS since the NRC 2004 report *Opportunities to Improve Airport Passenger Screening with Mass Spectrometry*. These are discussed in the following section.

Recent Developments in Mass Spectrometry

Ambient Ionization

Ambient ionization refers to a family of methods developed since 2004, in which samples are ionized in their native environment and original physical state without needing to be prepared by transferring analyte ions from near the surface of the sample into the vacuum system of the mass spectrometer. Several dozen ambient ionization methods have been described and reviewed in the recent literature.[4] The methods can be divided into those based on sprays of charged droplets and those based on plasmas or lasers. Desorption of material from the sample surface and ionization of that material are the two operations common to all methods. In some cases, such a desorption electrospray ionization (DESI)[5] and direct analysis in real time (DART),[6] the desorption and ionization steps are achieved by a single agent (charged droplets in DESI, gaseous metastable atoms and ions in DART). In other cases, independent methods are used to effect these two steps, as in laser ablation electrospray ionization,[7] in which a laser is used for desorption, and an electrospray to ionize the desorbed molecules in the gas phase. Some of these recent developments are illustrated in Figure C-1.

---

[4] R.G. Cooks, Z. Ouyang, Z. Takats, and J.M. Wiseman, Ambient mass spectrometry, Science 311:1566-1570, 2006; G. Van Berkel, Established and emerging atmospheric pressure surface sampling/ionization techniques for mass spectrometry, Journal of Mass Spectrometry 43:1161-1180, 2008.

[5] Z. Takats, J.M. Wiseman, B. Gologan and R.G. Cooks, Mass spectrometry sampling under ambient conditions with desorption electrospray ionization, Science 306:471-473, 2004.

[6] R.B. Cody, J.A. Laramee, and H.D. Durst, Versatile new ion source for the analysis of materials in open air under ambient conditions, Analytical Chemistry 77:2297-2302, 2005.

[7] P. Nemes and A. Vertes, Laser ablation electrospray ionization for atmospheric pressure, in vivo, and imaging mass spectrometry, Analytical Chemistry 79:8098-8106, 2007.

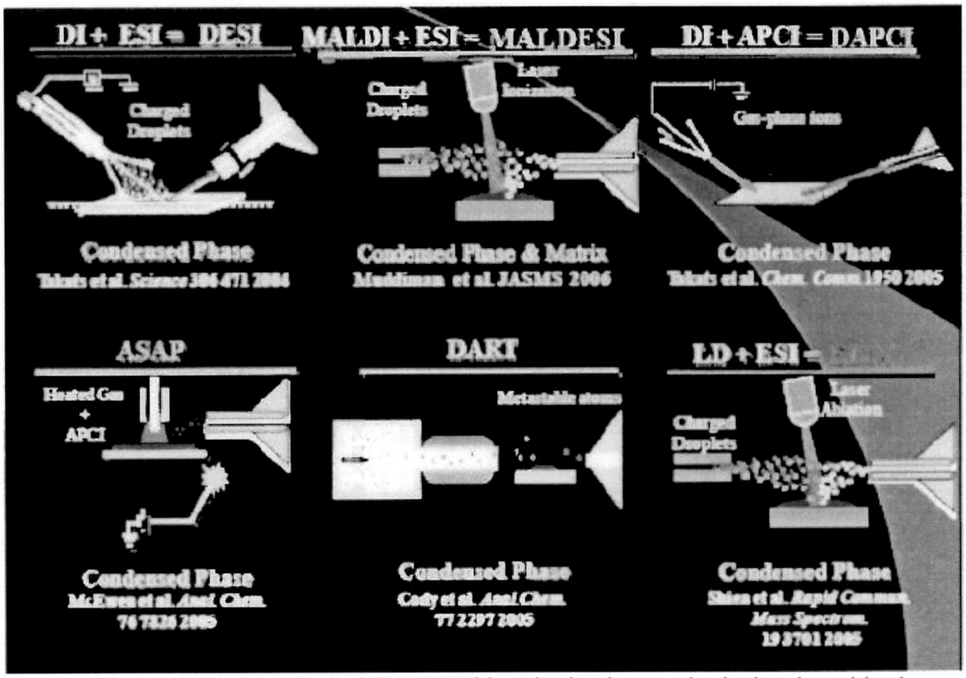

FIGURE C-1 Some of the several dozen ambient ionization methods developed in the past few years. SOURCE: Courtesy of R.G. Cooks, Purdue University, as described in the following: *DESI:* Adapted from Z. Takáts, J.M. Wiseman, B. Gologan, and R.G. Cooks, Mass spectrometry sampling under ambient conditions with desorption electrospray ionization, *Science* 306(5695):471-473, 2004; *ELDI:* J. Shiea, M.-Z. Huang, H.-J. HSu, C.-Y. Lee, C.-H. Yuan, I. Beech, and J. Sunner, Electrospray-assisted laser desorption/ionization mass spectrometry for direct ambient analysis of solids, Rapid Communications in Mass Spectrometry 19:3701-3704, 2005; *MALDESI:* J.S. Sampson, A.M. Hawkridge, and D.C. Muddiman, Generation and detection of multiply-charged peptides and proteins by matrix-assisted laser desorption electrospray ionization (MALDESI) Fourier transform ion cyclotron resonance mass spectrometry, Journal of the American Society for Mass Spectrometry 17(12):1712-1716; *EDI:* K. Hiraoka K. Mori, and D. Asakawa, Fundamental aspects of electrospray droplet impact/SIMS, Journal of Mass Spectrometry 41(7):894-902, 2006; *AP-MALDI:* V.M. Doroshenko, V.V Laiko, N.I. Taranenko, V.D. Berkout, and H.S. Lee, Recent developments in atmospheric pressure MALDI mass spectrometry, International Journal of Mass Spectrometry 221(1):39-58, 2002; *DART:* R.B. Cody, J.A. Laramée, and H.D. Durst, Versatile new ion source for the analysis of materials in open air under ambient conditions, Analytical Chemistry 77(8):2297-2302, 2005.

Among the recent plasma-based methods are those that are based on discharge barrier desorption ionization (DBDI),[8] a method that produces low-power, stable radio frequency plasmas in air or a noble gas. The low-temperature plasma (LTP) probe is a recent version of non-thermal (non-equilibrium) plasma.[9] The characteristic feature of this probe configuration is that it allows the plasma direct access to the sample's surface and near surface. Gas flows are minimal, and total power required is very low (around 3 W). Voltages used are in the kV range and frequencies in the kHz range. The method shows promise as an ionization method complementary to DESI: It ionizes many small molecules, including explosives, and generates mass spectra characterized by abundant molecular ions from which molecular weights are obtained and from which MS/MS spectra can be recorded showing characteristic fragment ions for compound identification.

[8] N. Na, M.X. Zhao, S.C. Zhang, C. Yang, and X. Zhang, Development of a dielectric barrier discharge ion source for ambient mass spectrometry, Journal of the American Society for Mass Spectrometry 18:1859-1862, 2007.

[9] J.D. Harper, N.A. Charipar, C.C. Mulligan, X. Zhang, R.G. Cooks, and Z. Ouyang, Low-temperature plasma probe for ambient desorption ionization, Analytical Chemistry 80:9097-9104, 2008.

A) B)

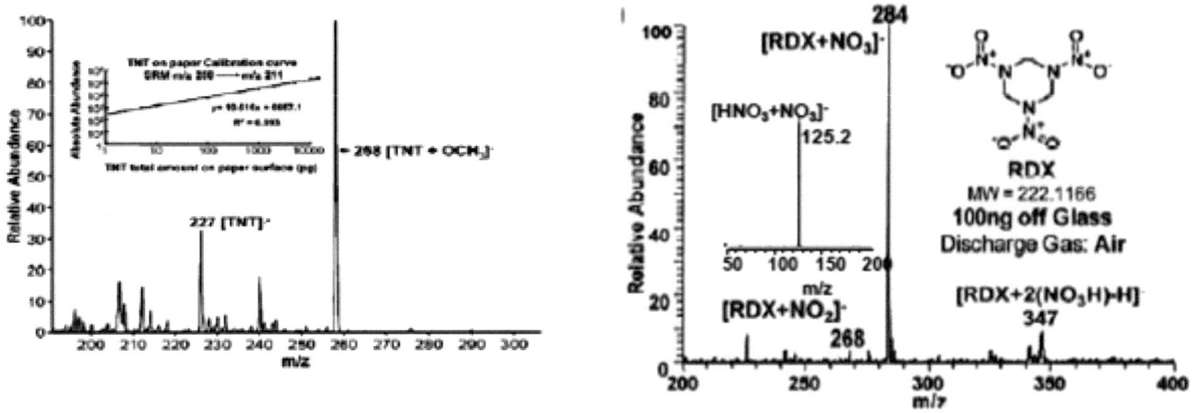

FIGURE C-2 (A) Negative ion DESI mass spectrum showing the formation of the Meisenheimer complex between TNT and the methoxide anion at m/z 258 when examining 10 pg of TNT deposited on paper in a total area of 1 cm^2. (B) negative ionization of 100 ng RDX by LTP; inset shows the nitrate cluster anion formation during LTP ionization in air. SOURCE: After I. Cotte-Rodriguez, Z. Takats, N. Talaty, H. Chen, and R.G. Cooks, Desorption electrospray ionization of explosives on surfaces: Sensitivity and selectivity enhancement by reactive desorption electrospray ionization, Analytical Chemistry 77:6755-6764, 2005.

The most extensive studies of ambient ionization of explosives have employed DESI. The more recent LTP method—like other plasma methods, including DBDI and plasma assisted-desorption ionization (PADI)[10]—is also attractive for explosives analysis so DESI and LTP are almost exclusively discussed in what follows. Both LTP[11] and DESI[12] allow nanogram amounts of explosives to be detected from ambient surfaces and characterized by highly specific MS/MS data in times on the order of a few seconds. Some typical data recorded using a commercial mass spectrometer are shown below for trinitrotoluene (TNT) and a peroxide explosive using DESI (Figure C-2).

The absence of a need for sample preparation in ambient ionization mass spectrometry means that high throughput can be achieved; most measurements take only a few seconds, including the time to confirm a compound seen in the mass spectrum as a characteristic ion through its MS/MS spectrum. Moreover, the low-impact nature of DESI and other methods of ambient ionization means that the mass spectra are dominated by intact molecular ions.

The specific identification of materials as particular chemical entities is arguably more important and more difficult than achieving the speed and sensitivity necessary for airport security detection purposes. This is in fact widely recognized as the main problem with ion mobility, which is fast and sensitive but not highly specific. In MS experiments, additional specificity is easily provided by MS/MS and in larger instruments by high-resolution measurements that give molecular formulas. Specificity can be increased further by another simple experiment, a "reactive" version of ambient ionization. These

[10] L.V. Ratcliffe, F.J. M. Rutten, D.A. Barrett, T. Whitmore, D. Seymour, C. Greenwood, Y. Aranda-Gonzalvo, S. Robnison, and M. McCoustra, Surface analysis under ambient conditions using plasma-assisted desorption/ionization mass spectrometry, Analytical Chemistry 79:6094, 2007.

[11] J.D. Harper, N.A. Charipar, C.C. Mulligan, X. Zhang, R.G. Cooks, and Z. Ouyang, Low-temperature plasma probe for ambient desorption ionization, Analytical Chemistry 80:9097-9104, 2008.

[12] I. Cotte-Rodriguez, Z. Takats, N. Talaty, H. Chen, and R.G. Cooks, Desorption electrospray ionization of explosives on surfaces: Sensitivity and selectivity enhancement by reactive desorption electrospray ionization, Analytical Chemistry 77:6755-6764, 2005; Z. Takats, I. Cotte-Rodriguez, N. Talaty, H.W. Chen, and R.G. Cooks, Direct, trace level detection of explosives on ambient surfaces by desorption electrospray ionization mass spectrometry, Chemical Communications 1950-1952, 2005.

experiments are done by simply adding a chemical reagent to the spray solution (DESI) or the support gas (LTP). For example, betaine aldehyde (BA) gives characteristic adducts with TNT during the DESI process when BA is incorporated in the spray solvent (Figure C-3).

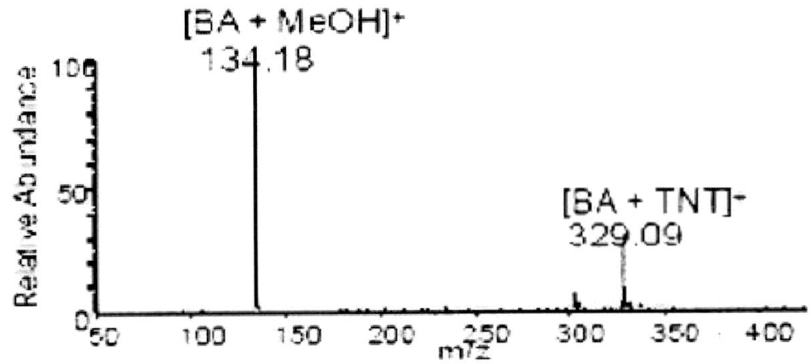

FIGURE C-3 Positive mode electrospray ionization mass spectrum acquired by spraying TNT sample with betaine aldehyde (BA) in methanol water. SOURCE: Data from Chumping Wu and R.G. Cooks.

Handheld Mass Spectrometers

The miniaturization of mass spectrometers is moving forward swiftly.[13] Some of the work deals only with the actual mass analyzer, but full, autonomous miniature MS systems are also in operation. Table C-1 summarizes Ouyang and Cooks'[14] information on the state-of-art in miniature mass spectrometers. Note that MS/MS capabilities are highly desirable for trace mixture analysis of explosives. In addition, ambient ionization is needed for rapid analysis. Both the DESI and LTP ionization techniques have been implemented on portable ion-trap mass spectrometers. Pulsed ion introduction—discontinuous atmospheric pressure introduction (DAPI)[15]—is essential for performing atmospheric pressure ionization, including ambient ionization because of the small pump sizes.

Negative ions are detected by incorporation of a conversion dynode in conjunction with a channeltron electron multiplier detector. Current efforts focus on optimization of ion transport in these experiments, which remain inefficient in spite of ability to detect subnanogram amounts of explosives using benchtop instruments. Most losses are in the atmospheric pressure region and involve failure to efficiently transport the atmospheric pressure ions/ionized droplets into the mass spectrometer.

[13] Z. Ouyang and R.G. Cooks, Miniature cylindrical ion trap mass spectrometer, Analytical Chemistry 74(24):6145-6153, 2002.

[14] Zheng Ouyang and R. Graham Cooks, Miniature mass spectrometers, Annual Review of Analytical Chemistry 2:187-214, 2009.

[15] L. Gao, R.G. Cooks, and Z. Ouyang, Breaking the pumping speed barrier in mass spectrometry: discontinuous atmospheric pressure interface, Analytical Chemistry 80:4026-4032, 2008.

TABLE C-1 Self-Sustainable Portable MS Systems

Instrument	Developer	Weight (kg)	Power (W)	Mass Analyzer	MS/MS	Sampling/ Ionization[a]	Mass Range/ Resolution
Mini 10/ Mini 11	Purdue University	10/4	70/30	Rectilinear ion trap	Yes	MIMS, direct leak, GDEI, APCI, ESI, DESI, LTP	m/z 550, R = 550; m/z 2,000, R = 100
ChemCube	Microsaic Systems	14	50	Quadruple mass filter	No	SPME, EI	m/z 400, R = 100
Guardion-7	Torion Technology	11	75	Toroidal ion trap	Yes	SPME, Mini GC EI	m/z 500, R = 500
Suitcase TOF	Johns Hopkins Applied Physics Lab	N/A	N/A	Time of flight	No	MALDI	m/z 70,000, R = 70
Griffin 600	Griffin Analytical	15	N/A	Cylindrical ion trap	Yes	SPME, MIMS, EI	m/z 425, R = 300
Ion-Camera	O-I-Analytical	18	75	Mattauch-Herzog sector	No	Direct gas leak, EI	m/z 300, R = 300

[a] MIMS membrane introduction mass spectrometry; GDEI glow discharge electron impact ionization; APCI atmospheric-pressure chemical ionization; ESI electrospray ionization; SPME solid-phase microextraction.
SOURCE: Adapted from Z. Ouyang, and R.G. Cooks, Miniature mass spectrometers, Annual Review of Analytical Chemistry 2:187-214, 2009.

The issue of large area detection[16] has been addressed also in benchtop instruments but not in miniature MS instruments. Larger areas (several hundred square centimeters) are accessible by LTP methods using several plasma probes. Comparable DESI experiments involve large amounts of solvent and are more awkward to implement.

Stand-off detection experiments have been surprisingly effective. Ions are transported back to the mass spectrometer over distances of several meters in both DESI and LTP experiments. Signals fall over several orders of magnitude in these experiments, but chemical noise falls faster, and high-quality explosives data can be recorded on benchtop lab instruments for samples of a few nanograms. Supplementary pumping greatly improves performance. The combination of large area detection and stand-off detection will be difficult to achieve, and neither kind of detection is easily achieved in miniature instruments.

The power of handheld mass spectrometers is illustrated by the fact that they can be used to perform multiple-stage MS experiments. Such experiments add great specificity to identifications made by MS and usually require little extra time to perform. The relevant capabilities of the combination of miniature mass spectrometer and ambient ionization are illustrated in Figure C-4, which shows LTP and DESI spectra taken on small amounts of sample with these methods.

[16] Santosh Soparawalla, Gary A. Salazar, Ewa Sokol, Richard H. Perry, and R. Graham Cooks, Trace detection of explosives distributed over large areas using mass transfer and ambient ionization mass spectrometry, Analyst 135:1953-1960, 2010.

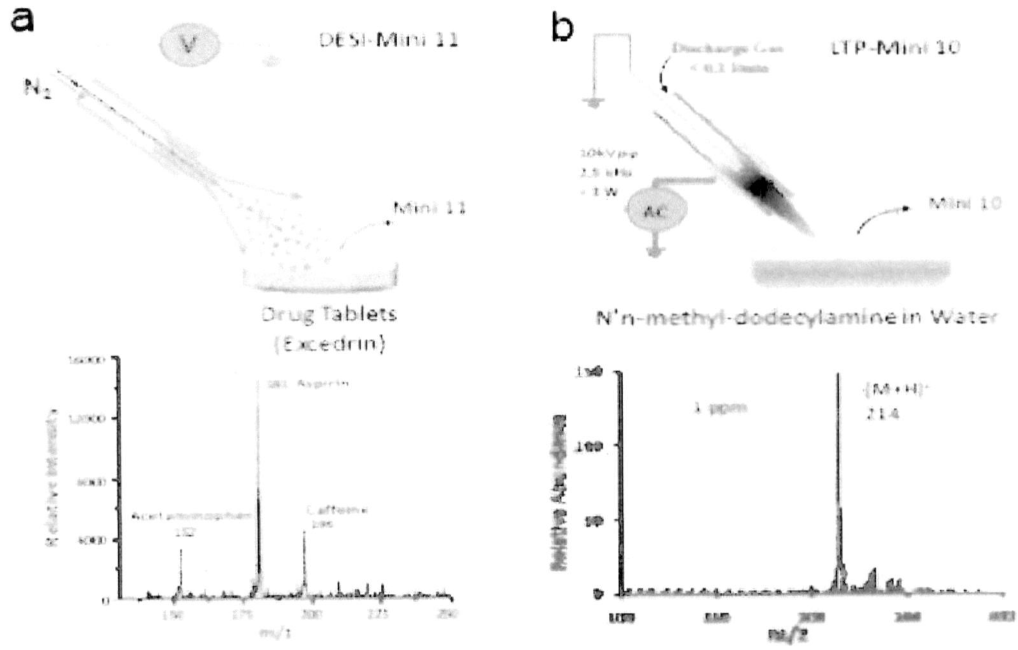

FIGURE C-4 Use of DESI and LTP ambient ionization methods to detect compounds in ordinary materials.

ADVANTAGES OF MASS SPECTROMETRY IN TRACE EXPLOSIVES ANALYSIS

It is clear from the data collected here that a new generation of mass spectrometers has already emerged in research labs with capabilities that are potentially applicable to airport screening of baggage and passengers. Among the favorable characteristics of these instruments are their small size and the highly capable ambient ionization methods, which are rapid and sensitive and yet give a great deal of specific information on the chemical nature of a particular sample, including information on the presence of traces of explosives on surfaces. Other characteristics, like the ability to perform stand-off MS detection, the ability to add specificity by "reactive" ionization methods, the ability to quantify, and the ability to extend the methodology to surfaces with large areas are less well developed but are under study.

While the direct analysis of surfaces for explosives is the simplest and most desirable implementation of these capabilities, the swabbing of surfaces with the standard "swiffer" wipes used in security lines today could also be used in secondary passenger and baggage screening, and it could be done by DESI MS with much greater chemical specificity (fewer false positives) than when done by ion nobility.

D

Statistical Approaches to Reducing the Probability of False Alarms While Improving the Probability of Detection

This section suggests two statistical approaches to improve the detection probability and reduce the probability of a false positive. The first is based on some very basic statistical concepts of testing simple versus simple[1] hypothesis using the Neyman-Pearson (NP) lemma. The second is an evidential approach. The idea behind this approach is quite simple. Currently a bag is passed through the CT scanner once. After this single pass, the bag is either cleared or is sent to a human screener. The decision is based on the automatic feature recognition program. If the bag is sent to a human screener, this person looks at the CT scanner image and either clears it or sends it for inspection by hand. Another idea is to send the bag through the CT scanner more than once and depending on the number of times the bag is flagged as a threat, it is either cleared or sent to a human screener. The main assumption is that each time the bag passes through the CT scanner, it provides a different scan. This is reasonable because the bag almost certainly will get positioned somewhat differently at each pass because of the bumps on the conveyor belt. The CT scanner does not know it is the same bag that it is scanning, so the scans are independent of one another.

In the following, the false positives are the number of bags sent to the human screener that are "non-threat" bags. Detection probability is the probability of detecting a threat when it exists.

NOTATION

The following notation is used:

- P (A bag is declared a "threat" by the CT scanner | the bag is a true threat) = δ. This is the probability of detecting a threat by the CT scanner in one scan.
- P (A bag is declared a "threat" by the CT scanner | the bag is not a true threat) = α. This is the probability of *falsely* detecting a threat by the CT scanner in one scan.

Thus,

True State	Machine Output	
	Declare threat (1)	Not declare threat (0)
Threat	δ	$1-\delta$
No threat	α	$1-\alpha$

These probabilities are estimable from the experiments that TSA currently conducts.

[1] When the true state of nature is dichotomous (i.e., threat or no threat), in the statistical testing of hypothesis terminology, testing for which state the data support the most is called a simple versus simple hypothesis testing problem. See also G. Casella and R.L. Berger, Statistical Inference, 2nd edition, Duxbury Press, Pacific Grove, Calif., 2002, Chapter 8.

PROBLEM FORMULATION

The problem can be formulated as a statistical testing of hypotheses.

Hypothesis 1: The bag is not a threat.
Hypothesis 2: The bag is a threat.
Given: *Y*, the number of times out of *N* the bag is declared a threat by the CT scanner

Statistical Model

It is obvious that

(a) $Y \sim Binomial\ (N, \alpha)$ under Hypothesis 1.
(b) $Y \sim Binomial\ (N, \delta)$ uner Hypothesis 2.

Decision Process

Pass the bag N number of times. If q (or more) out of N tests are positive, send the bag to the human screeners and human inspectors.

Relevant Probabilities

1. Probability of declaring a bag to be a threat when the threat exists (correct detection):

$$P(Y \geq q \mid Hypothesis\ 2) = \sum_{i=q}^{N} \binom{N}{i} \delta^i (1-\delta)^{N-i} = P_D$$

2. Probability of declaring a bag to be a threat when it is not a threat (false alarm):

$$P(Y \geq q \mid Hypothesis\ 1) = \sum_{i=q}^{N} \binom{N}{i} \alpha^i (1-\alpha)^{N-i} = P_F$$

For the following calculations, assume that $\alpha = 0.2, \delta = 0.9$. Any other appropriate values may be substituted in the above formulas to obtain the relevant probabilities (Table D-1). A typical entry in the table is read as: Suppose the decision rule is such that a bag is declared a threat if it tests positive at least three times out of the total of five scans. Such a decision rule will detect the threat correctly 99.14 percent of the time and will give a false positive alarm 5.79 percent of the time.

Policy makers can decide the appropriate values for N and q based on the values of δ and α and the desired probabilities of detection and the false alarm. Given the often cited approximate number of annual savings of $25 million per percentage point drop in the false alarm rate, this simple scheme represents a potential saving of $375 million per year.

TABLE D-1 Probabilities for Some Combinations of N and q

N	q	Correct Detection	False Alarm
2	1	0.99	0.36
	2	0.81	0.04
3	1	0.999	0.488
	2	0.972	0.104
	3	0.729	0.008
4	1	0.9999	0.5904
	2	0.9963	0.1808
	3	0.9477	0.0272
	4	0.6561	0.0016
5	1	0.99999	0.67232
	2	0.99954	0.26272
	3	0.99144	0.05792
	4	0.91854	0.00672
	5	0.59049	0.00032

Evidential Approach

In the NP approach, we answered the question, given these data, what do I do? A somewhat different question may also be asked: Given these data, what strength of evidence do we have for the hypothesis "This bag is not a threat" vis-à-vis "This bag is a threat"? Using the law of the likelihood[2] this is given by the likelihood ratio

Strength of evidence for "no threat": $\frac{P(q \text{ flags out of N trial } | \text{No threat})}{P(q \text{ flags out of N trial } | \text{Threat})} = \frac{\alpha^q \, (1-\alpha)^{N-q}}{\delta^q \, (1-\delta)^{N-q}}.$

The policy makers have to decide at which level of strength of evidence for "no threat" the bag may be cleared. If the strength of evidence for no threat is below that level, the bag will be sent for hand inspection. For the sake of illustration, suppose we say that if the strength of evidence for no threat is larger than 4, the bag will be cleared; otherwise it will be sent to for hand inspection. Under the "no threat" hypothesis, we can compute how often we would send the bag for hand inspection (probability of false positives) and under the "threat" hypothesis, how often would we clear the bag (probability of misleading evidence). Table D-2 was computed assuming a cut-off level of 4, where number 4 implies that the bag is four times more likely to not be a threat than to be a threat.

---

[2] Ian Hacking, The Logic of Statistical Inference, Cambridge University Press, Cambridge, U.K., 1965; Richard M. Royall, Statistical Evidence: A Likelihood Paradigm, Capman and Hall, New York, 1997; Mark L. Taper and Subhash R. LeLe, eds., The Nature of Scientific Evidence: Statistical, Philosophical, and Empirical Considerations University of Chicago Press, Chicago, Ill., 2004.

TABLE D-2 Threat Cut-off at 4

Number of trials	If the number of alarms is ≤ to this number, clear the bag[a]	Probability of clearing a bag when it is not a threat	Probability of not clearing a bag when it is not a threat (false positive)	Probability of clearing a bag when it is a threat (false negative)	Probability of detecting a threat when the threat exists
1	0	0.8	0.2	0.1	0.9
2	0	0.64	0.36	0.01	0.99
3	1	0.896	0.104	0.028	0.972
4	1	0.819	0.181	0.0037	0.9963
5	2	0.942	0.058	0.009	0.991

NOTE: A typical entry in this table can be read as "If we conduct five trials and fewer than three of them are positive, then we clear the bag." If we follow this rule, the probability of a false positive is 0.058 and the probability of detecting the threat is 0.991.
[a] This number is a function of (α,δ,K).

The decision rule will depend on the choice of K. The difference between the evidential approach and the NP approach is this: In the evidential paradigm the cut-off point is determined a priori and the error probabilities are calculated afterwards, whereas in the NP approach, the error probabilities are fixed a priori and the cut-off points are determined afterwards. In this particular situation, the author does not see any difference between following the NP approach or the evidential approach.

Incorporating Perception of Threat

The methodology described above does not incorporate the perception of threat. It was based simply on the data observed for a particular checked bag. If the perception of threat can be quantified, we can address the question, Given these data, how do I change my beliefs? The prior belief or perception of threat level can be incorporated into the above setup in the following fashion:

Let *P*(Threat) = π denote the perceived probability of threat. This represents our prior belief that the bag is a "threat" without having observed any data on a particular bag. This is equivalent to the epidemiological concept of "prevalence of a disease in the population." Now, having observed that the bag was flagged as a threat q times out of N passes, we want to know, in the light of these data, how we change our perception about the threat that this particular bag poses. This can be calculated quite easily using standard conditional probability calculations[3] as follows:

$$P(Threat \mid q \text{ out of N tests are positive}) = \frac{\delta^q (1-\delta)^{N-q}\pi}{\delta^q (1-\delta)^{N-q}\pi + \alpha^q (1-\alpha)^{N-q}(1-\pi)}.$$

Notice that this probability depends strongly on the value of π. Computing the probabilities of false negatives and correct detection depends on the specification of π and the cut-off point above which we declare a bag to be a threat. Hence it is not possible to present a comparison that captures perception of threat in relation to either the NP or the evidential approach.

[3] G. Casella and R.L. Berger, Statistical Inference, 2nd edition, Duxbury Press, Calif., 2002.

Comments

Implementation of this scheme will need to take into account the costs involved in scanning the bags repeatedly and tracking the number of times they are declared "positive" by the CT scanner. This may be facilitated easily by the RFID tag on each bag. The author does not feel qualified to comment on the feasibility of this aspect.

The conveyor belts that handle the bags will need to be redesigned to allow the bags to be scanned repeatedly and in such a manner that at each pass the position of the bag is perturbed to some extent. This would seem to be a manageable mechanical engineering problem. However, again the author declares that he is not qualified to comment on the feasibility and the cost of such changes.

This method can be easily modified if K different tests (machines) are used.

This can also be done in a sequential fashion where N is random and at each scan the decision is made whether to pass the bag or to send it to the human inspectors or to run another test. In the author's opinion, sequential design is logistically more complicated than the fixed N design described above.

The information from multiple scans, if made available to the human screener, might further reduce the number of bags that are sent for hand inspection.

DISCUSSION

Current technologies for scanning the checked baggage do a very respectable job. However, there are limits as to how much CT scanning technology and the feature detection algorithms can increase the probability of detection. The repeated scanning approach takes the current technology and significantly increases the probability of detection and decreases the probability of false alarm without requiring significant technological breakthroughs.

E

Statement of Task

An ad hoc committee will examine the technology of current aviation-security explosive-detection systems (EDSs) and the false positives produced by this equipment. In assessing methods to reduce the EDS false-alarm rate, the committee will:

1. Examine and evaluate the causes of false positives in aviation explosive-detection systems, including considering the role of equipment design standards that rely on the fusion of explosive density measurement, total mass, and shape effects.
2. Assess the impact false positive resolution has on personnel and resource allocation.
3. Make recommendations on mitigating false positives without increasing false negatives, considering both technology and personnel approaches and related short- and long-term research. The committee recommendations will also bear in mind any risk of increased missed detection.

F

Acronyms and Definitions of Selected Terms

ACRONYMS

ATR	automated threat recognition
BA	betaine aldehyde
BHS	baggage-handling system
BIR	baggage-inspection room
BVS	baggage-viewing station
CC	control computer
CT	computed tomography
DAPI	discontinuous atmospheric pressure introduction
DART	direct analysis in real time
DAS	data acquisition system
DBDI	desorption ionization
DESI	desorption electrospray ionization
DHS	Department of Homeland Security
DHS S&T	Department of Homeland Security's Science and Technology Directorate
DICOM	Digital Imaging and Communications in Medicine
DICOS	Digital Imaging and Communication in Security
DOD	Department of Defense
DOT	Department of Transportation
EDS	explosive detection system
ETD	explosive trace detection
FAA	Federal Aviation Administration
FAT	factory acceptance test
FBP	filtered back-projection
FDA	Food and Drug Administration
FDRS	field data reporting system
HU	Hounsfield unit
HVPS	high voltage power supply
ID	identification
IED	improvised explosive device
IMS	ion mobility spectrometry
IR&D	internal research and development

ITRS	International Technology Roadmap for Semiconductors
LTP	low temperature plasma
MS	mass spectrometry
NEMA	National Electrical Manufacturers Association
NRC	National Research Council
O&M	operations and maintenance
ODT	ordnance disposal team
OEM	original equipment manufacturer
OSARP	on-screen alarm-resolution protocol
OSR	on-screen resolution
PBL	performance-based logistics
PD	probability of detection
PFA	probability of false alarm
PVS	primary viewing station
QRA	quantitative risk assessment
RDT&E	research, development, testing, and evaluation
ROC	receiver operator characteristic
SAT	site acceptance test
SOP	standard operating procedure
SSI	sensitive security information
SSR	system status rate
SVS	secondary viewing station
TSA	Transportation Security Administration
TSL	Transportation Security Laboratory
TSO	transportation security officer
XRD	x-ray diffraction

DEFINITIONS OF SELECTED TERMS

alarm: A portion of a bag that is a potential threat as determined by the automated threat recognition algorithm.

bag: Item scanned by the explosive detection system. This is usually a piece of baggage, but could be items in bins or small pieces of cargo.

clearing: The process of the automated threat recognition (ATR) algorithm's indicating that a threat is not present in a bag or that the decision of the ATR algorithm is overridden by secondary inspection.

explosive detection system: Used for checked-baggage screening at airports: computed tomography-based device for interrogating a bag; composed of a computed tomography scanner, automated threat recognition algorithm, a workstation, and a control computer.

false alarm: Sometimes called a false positive; the automated threat-recognition algorithm signals an alarm, but no threat is present in the bag being screened.

mis-track: A bag that cannot be tracked by the baggage-handling system.

on-screen alarm-resolution protocol: TSA process by which a human screener resolves an alarm based on the image from the scanner

shield: The condition that occurs when the explosive detection system cannot view a portion of a bag because the x-ray beam is extinguished by the presence of clutter.

threat: A portion of a bag that is a potential threat as determined by the automated threat recognition algorithm.

transportation security officer: Operator of the baggage-viewing station and worker in the baggage-inspection room.